Introduction To Wireless Systems

Technologies, Systems, Services and Market Growth

Lawrence Harte

2nd Edition

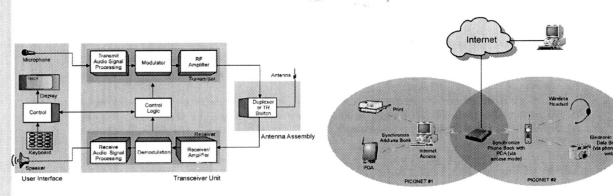

Mobile Radio Functions

Bluetooth System

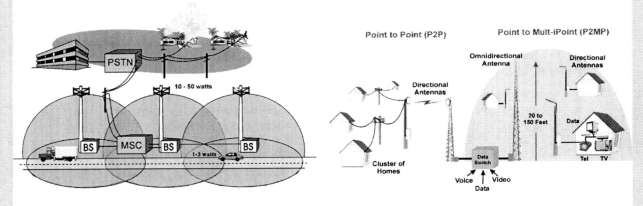

Mobile Telephone System

WiMax System

Excerpted From:

Wireless Systems

With Updated Information

ALTHOS

ALTHOS Publishing

ALTHOS Publishing

ALTHOS electronic books (ebooks) and images are available for use in educational, promotional materials, training programs, and other uses. For more information about using ALTHOS ebooks and images, please contact April Wiblitzhouser awiblitzhouser@Althos.com or (919) 557-2260.

Terms of Use

About the Author

Mr. Harte is the president of Althos, an expert information provider whom researches, trains, and publishes on technology and business industries. He has over 29 years of technology analysis, development, implementation, and business management experience. Mr. Harte has worked for leading companies including Ericsson/General Electric, Audiovox/Toshiba and Westinghouse and has consulted for hundreds of other companies. Mr. Harte continually researches, analyzes, and tests new communication technologies, applications, and services. He has authored over 60 books on telecommunications technologies and business systems covering topics such as mobile telephone systems, data communications, voice over data networks, broadband, prepaid services, billing systems, sales, and Internet marketing. Mr. Harte holds many degrees and certificates including an Executive MBA from Wake Forest University (1995) and a BSET from the University of the State of New York, (1990).

Table of Contents

Introduction to Wireless Systems

Wireless systems are the combination of equipment, protocols and transmission lines that are used to provide wireless communication services. Wireless communication systems can transfer a mix of voice, data or video information.

Because wireless communication systems transport information, the types of services these systems offer vary dependent on the applications they are designed to support. Wireless applications are created to produce benefits for their users. Successful wireless communication applications include mobile telephones, land mobile radio, paging, wireless data, fixed wireless, remote control, point-to-point communications, broadcast radio, wireless broadband data and television.

Radio Frequency (RF)

Radio is the transfer of information signals through the air or on transmission lines by the means of electromagnetic waves. These electromagnetic waves repeat their cycle in a frequency range of approximately 150 kHz to 300GHz.

The radio frequency spectrum is divided into frequency bands that are authorized for use in specific geographic regions. Globally, the International Telecommunications Union (ITU) specifies the typical use for radio frequency bands. Within each country, government agencies create and enforce the rules for which specific types of systems and services are used in specific frequency bands and which companies will be able (will be licensed) to own and operate these systems.

Licensing

The national government is responsible for dividing the available frequency bands for licensing to users and regulates what the frequencies may be used for. The legal right-to-use of this public resource is controlled by rules and licensing of very specific frequencies, a range of frequencies or a block of subdivided channels at a given frequency or frequency range.

For example, the frequencies allocated for FM radio must be used for the purpose licensed; that is a combination of music or news and public information. FM radio stations are not licensed to broadcast a secret "Morse-code" to a following of undercover militia! Neither can a "Paging Service" use one or all of their frequency channels to broadcast radio. However, with the recent deregulation of telecommunications services, wireless service providers are now permitted to offer many new types of services provided they can fulfill their basic licensing requirements.

To prevent unwanted interference from radio devices, unauthorized use of transmitting energy or information on our public airwaves according to publicly published rules or licenses may violate governmental laws. Such transmissions are subject to prosecution or suspension of the radio operator's license.

Unlicensed Frequency Bands

Unlicensed frequency bands are a range of frequencies that can be used by any product or person provided the transmission conforms to transmission characteristics defined by the appropriate regulatory agency.

Unlicensed frequency bands are regulated and manufacturers of products that operate in these frequency bands must conform to these regulations.

These regulations define how these devices may operate to ensure that they can independently operate with minimal interference to each other. Examples of unlicensed devices include wireless local area networks, keychain remotes for cars, wireless baby monitors and radio operated garage door operators.

Frequency Allocation

Frequency allocation is the amount of radio spectrum (frequency bands) that is assigned (allocated) by a regulatory agency for use by specific types of users or operators for defined types of services. The selection of the assigned frequency bands is determined by a variety of factors including the availability of radio channel frequencies, the radio propagation characteristics and the needs for wireless communication services.

Because most of the frequencies have already been assigned to licensees, frequency reallocation is used to make radio frequencies available for assignment. Frequency reallocation is the revocation and re-assignment of frequency licenses to users or operators to allow for new users or operators to use an existing frequency band. An example of frequency reallocation is the reassignment of licenses for fixed microwave transmission services in the 2 GHz frequency band to higher frequencies to allow in the use of the 2 GHz frequency for mobile telephone services. As part of this reallocation process, the new licensees of the 2 GHz frequency band were required to pay the costs of converting existing license holders to the new frequency bands.

RF Channels and Bandwidth

An RF channel is a communication link that use radio signals to transfer information between two (or more) points. To transfer this information, a radio wave (typically called a radio carrier) is modulated (modified) within an authorized frequency band to carry the information. The modulation of the radio wave forces the radio frequency to shift above and below the reference (center) frequency. Typically, the more the modification of frequency, the more information can be carried on the radio wave. This results in RF channels typically defined by their frequency and bandwidth allocation.

Bandwidth allocation is the frequency width of a radio channel in Hertz (high and low frequency limits) that can be modulated (changed) to transfer information (voice or data signals). The amount type of information being sent determines the amount of bandwidth used and the method of modulation used to impose the information on the radio signal.

A government regulation agency (the FCC in the United States) defines a total frequency range (upper and lower frequency limits) that a radio service provider can use to transmit information. In some systems (such as AM or FM radio station broadcasting), this is a single radio channel. For other systems (such as cellular, PCS, or PCN), this is a range of frequencies that can be sub divided into smaller radio channels as determined by the radio carrier. When the allocated frequency range is further subdivided into smaller allowable bands, these subdivided areas are referred to as channels.

Overview

Wireless networks are composed of radios, radio towers or base stations, interconnection systems, and network management and information systems.

Radios

Radios may be fixed in location (such as a television) or may be mobile (such as a cellular or PCS telephone). Some radios may only communicate in one direction (typically a receiver) or may have two-way capability. When a single radio has both a transmitter and receiver contained in the same unit, it is called a transceiver.

Figure 1.1 shows a block diagram of a mobile radio transceiver. In this diagram, sound is converted to an electrical signal by a microphone. The audio signal is processed (filtered and adjusted) and is sent to a modulator. The modulator creates a modulated RF signal using the audio signal. The modulated signal is supplied to an RF amplifier that increases the level of the RF signal and supplies it to the antenna for radio transmission. This mobile radio simultaneously receives another RF signal on a different frequency to allow the listening of the other person while talking. The received RF signal is then boosted by the receiver to a level acceptable for the demodulator assembly. The demodulator extracts the audio signal and the audio signal is amplified so it can create sound from the speaker.

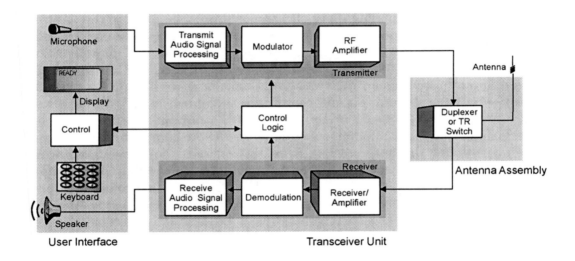

Figure 1.1, Mobile Radio Block Diagram

Radio Towers and Transmitter Equipment

Radio towers are poles, guided towers, or free standing constructed grids that raise one or more antennas to a height that increases the range of a transmitted signal. Radio towers can vary in height from about 20 feet to more than 300 feet. In general, the higher the antenna height (e.g. tower height), the greater the distance the radio signal can reach.

A single radio tower may host several antenna systems that include paging, microwave, or cellular systems. Radio towers are located strategically around the city to provide radio signal coverage to specific areas. At the base of the towers are electronic control rooms that contain the components to operate the radio portion of the communications system. When the radio equipment is located at the base of the radio tower, it is called a base station.

Radio towers and their associated radio equipment (e.g., base station) may include one or more antennas, transmitters, receivers (for two-way systems), system controllers, communication links, and power supplies. Transmitters provide the high level RF power that is supplied to the antenna. For broadcast systems, the amount of transmitter power can exceed 50,000 Watts. Receivers boost and demodulate incoming RF signals from mobile radios. If a base station contains receivers, it is typical to use one or more different antennas for the receivers. Controllers coordinate the overall operation of the base station and coordinate the alarm monitoring of electronic assemblies. Communication links allow a command location (such as a television studio or a telephone switching center) to control and exchange information with the base station. Base station radio equipment requires power supplies. Most base stations contain primary and backup power supplies. A battery typically maintains operation when primary power is interrupted. A generator may also be included to allow operation during extended power outages.

Figure 1.2 shows a typical radio base station block diagram that is used in a mobile telephone system. This diagram shows that the base station holds the radio transceiver (transmitter and receiver assemblies) that is part of the radio tower (cell site). This diagram also shows that one antenna is used for transmitting and two antennas are used for receiving (for improved reception). This base station also contains a backup battery that is maintained at full charge so radio communications will not be interrupted in the event AC power is lost.

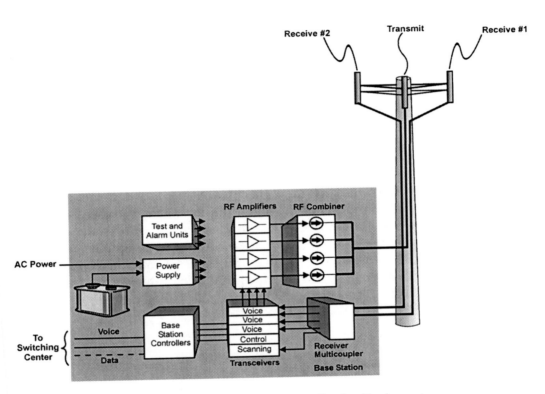

Figure 1.2, Radio Tower and Base Station Equipment

Switching Facilities

Switching is the process of connecting two (or more) points together. Switching may involve a single physical connection (such as a light switch) or it may involve the setup of multiple connections within a network through several communication devices. Switching facilities may be used to provide relatively long term connections (circuit connections) or for very short connections that can dynamically change (packet data connections). When switching systems are used in mobile communication systems, they are called mobile switching centers (MSC).

Mobile switching centers are used in two-way mobile communication systems to allow the connection of mobile telephones to other mobile telephones that are operating in the system or to connect mobile telephones to other networks such as the public switched telephone network (PSTN). The MSC processes requests for service from mobile radios (subscribers) and decides how to route the calls to other destinations.

Figure 1.3 shows that a MSC consists of controllers, switching assemblies, communications links, operator terminals, subscriber databases, and back-up energy sources. The controllers are computers which control communications and call processing. Controllers help the MSC to understand and create commands to and from the base stations. In addition to the main controller, secondary controllers devoted specifically to control of the cell sites (base stations) and to handling of the signaling messages between the MSC and the PTSN are also provided. A switching assembly routes voice connections from the cell sites to each other or to the public telephone network. Communications links between cell sites and the MSC may be copper wire, microwave, or fiber optic. An operator terminal allows operations, adminis-

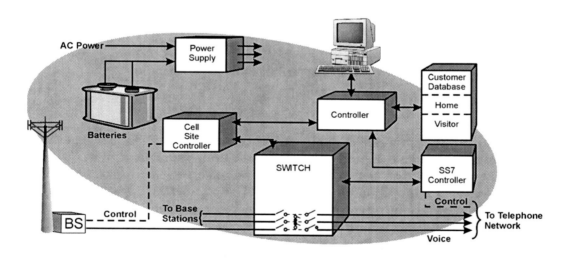

Figure 1.3, Wireless Switching System Block Diagram

tration and maintenance of the system. A subscriber database contains features the customer has requested along with billing records. Backup energy sources provide power when primary power is interrupted. As with the base station, the MSC has many standby duplicate circuits and backup power sources to allow system operation to be maintained when a failure occurs.

Interconnection to Other Networks

Wireless systems may be connected to other networks. Broadcast wireless systems are connected to media sources (such as audio or video programs) via satellite links while cellular networks may be interconnected to the public telephone network. Interconnection involves the physical and software connection of network equipment or communications systems to the facilities of another network such as the public telephone network. Government agencies such as the Federal Communications Commission (FCC) or Department of Communications (DOC) regulate interconnection of wireless systems to the public telephone networks to ensure fair and reliable operation.

Customer Databases

Customer databases are computer storage devices (typically a computer hard disk) that hold service authorization and feature preferences of customers. For wireless systems that allow the customer to operate in other territories, a home (local) database is used. Each wireless subscriber has a real-time user profile in the database that is typically called the home location register (HLR). The HLR identifies the current location of the mobile radio, the most likely place for the mobile to be, or the last location the subscriber was active. The MSC system controller uses this information to route calls to the appropriate radio tower for call completion. If the wireless user is not in a predetermined "home" range of the MSC, the mobile will register back through to the home signaling system to its home location register (HLR) for profile information.

When customers use the wireless services of systems outside of their home area, their information is transferred to a database in that system called the visitor location register (VLR). The VLR is part of a wireless network (typically cellular or PCS) that holds the subscription and other information about visiting subscribers that are authorized to use the wireless network.

System Security

In some wireless networks, access to system services requires validation of the customer's identity. These systems may use an authentication center (AUC) to store and process secret data to stop fraudulent calls or prohibit access to other paid for subscription services.

Wireless phones transmit some of their identification information over the public airwaves when they attempt to access the system. Thieves may try and intercept this information and copy (clone) the identification information that would allow them to make phone calls that would be billed to the other telephone. To prevent this unauthorized duplication of identification information, an authentication process can be used that uses secret keys to validate access information.

During the authentication process, code keys are created from secret codes that are stored in both the mobile radio and in the system. Along with basic identification information, these keys are exchanged during each system access attempt. The secret codes are not transmitted. Because the system and the mobile radio have the secret keys, both the mobile phone and the system can validate that the code information is correct. If the codes do not match, the system should not allow the call to be processed. New codes are created during each access attempt to prevent copying of the codes and immediately attempting access.

Market Growth

In 2006, approximately 1 in 2 people in the world were using mobile telephones. The percentage of people who use wireless devices (such as mobile telephones) in some countries had exceeded 100% (more than 1 mobile telephone per person). Some of the key market growth areas for wireless services include mobile telephone information services and wireless broadband service.

Mobile Telephone Service

By mid-2006, there were over 2.4 billion mobile telephone users in the world [1]. Some of the key drivers for continual growth include new services (Internet access), smaller and lower cost mobile telephones, and the availability of pre-paid wireless services.

Figure 1.4 shows the growth of mobile telephone customers throughout the world. This graph shows that the growth rate of the number of new mobile

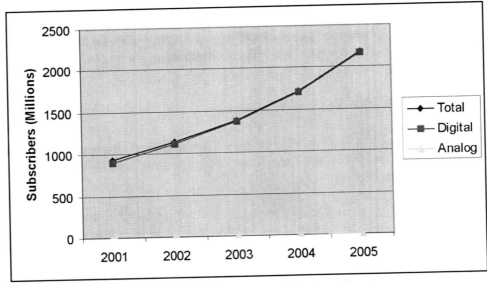

Figure 1.4, Mobile Telephone Wireless Growth

Source: GSM World

telephone subscribers in throughout the world is beginning to slow. This is a result of the high percentage of people that already have mobile telephone service. This graph also shows that a majority of mobile telephone subscribers use digital mobile telephone services.

Mobile Data Networks

With the demand for low cost wireless data communication solutions, paralleling interest in the Internet, (fueled by easy-to-use application software, its wide array of text, graphics, video and audio content,) wireless data market growth has increased substantially. The availability of data services over several types of wireless channels (cellular, two-way paging, packet data networks) is a critical factor in the overall market growth of wireless products and services.

To date, most wireless data applications are non-human in nature. These applications include remote meter reading, wireless parking meters, vending machines, and environmental concerns among others. Human access includes the ability to access data available on the Internet, private intranets, new services, and e-mail. The Internet, for example, is being used by businesses for building interactive branding via communication with customers, advertising products and services, publishing product specifications; and acting as a source for point-of-sale applications.

Wireless Broadband

Wireless broadband is the transfer of high-speed data communications via a wireless connection. Wireless broadband often refers to data transmission rates of 1 Mbps or higher.

Until the mid 2000s, many of the wireless broadband solutions used proprietary radio transmission technology. In the early 2000s, a standard wireless broadband technology was developed called worldwide interoperability for microwave access (WiMax). The availability of the WiMax industry standard has resulted in the availability of products at much lower cost. This is resulting in a shift of proprietary (company unique) solutions to WiMax systems.

Figure 1.5 shows the projection of growth for the wireless broadband industry. This diagram shows that until 2006, most wireless broadband systems used a proprietary (company unique) format. This diagram also shows that the wireless broadband industry is projected to grow from 3.5 million subscribers in 2006 to more than 87 million in 2012.

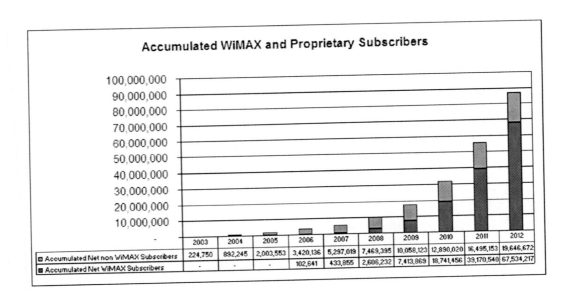

Accumulated WiMAX and Proprietary Subscribers

	2003	2004	2005	2006	2007	2008	2009	2010	2011	2012
Accumulated Net non WiMAX Subscribers	224,750	892,245	2,003,553	3,420,136	5,297,019	7,469,395	10,058,123	12,890,020	16,495,153	19,646,672
Accumulated Net WiMAX Subscribers	-	-	-	102,641	433,855	2,606,232	7,413,869	18,741,456	39,170,540	67,534,217

Figure 1.5, Wireless Broadband Market Growth
Source: Maravedis (www.maravedis-bwa.com)

Technologies

Key enabling technologies for wireless communication include digital modulation, data compression, and digital signal processing.

Digital Modulation

Digital modulation is the process of modifying the amplitude, frequency, or phase of a carrier signal using the discrete states (On and Off) of a digital signal.

When modulating a carrier signal using a digital information signal, this causes rapid changes to the carrier wave. These rapid changes result in the creation of other signals that are usually undesirable. As a result, digital modulation usually includes a process of adjusting the maximum rate of change of the input signal (rounding the digital signal edges) and filtering out some of the unwanted signals that are created during the transition.

Figure 1.6 shows different forms of digital modulation. This diagram shows ASK modulation that turns the carrier signal on and off with the digital signal. FSK modulation shifts the frequency of the carrier signal according to the on and off levels of the digital information signal. The phase shift modulator changes the phase of the carrier signal in accordance with the digital

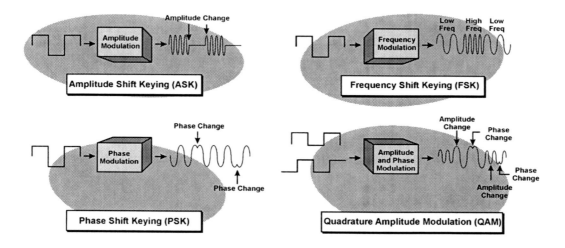

Figure 1.6, Digital Modulation

information signal. This diagram also shows that advanced forms of modulation such as QAM can combine amplitude and phase of digital signals.

Adaptive Modulation

Adaptive modulation is the process of dynamically adjusting the modulation type of a communication channel based on specific criteria (e.g. interference or data transmission rate).

In general, the more efficient (data transmission capacity) the modulation type, the more complex (precise) the modulation process is. The more precise the modulation process (smaller changes represent digital bits), the more sensitive the modulation is to distortion or interference. This usually means that as the data transmission rate increases, this increases the sensitivity to interference.

To help manage this process and ensure the maximum data transmission rate possible, wireless systems may be designed to automatically change their data modulation types and data transmission rates (Autorate) based on the ability of the channel to transfer data. Adaptive wireless systems will usually try to send information at the highest data transmission rate possible. If the data transmission rate cannot be maintained, the wireless systems will attempt to transmit at the next lower data transmission rate. Lower data transmission rates generally use a less complex (more robust) modulation type.

Information Compression

Information compression is the processing of information to a form that reduces the space or signal information (e.g. bandwidth) required for the transmission or storage of information. While it is possible to compress analog signals (e.g. such as signal companding and expanding), the information compression that is used in mobile communications is digital compression. Two key areas of digital compression that is used in wireless systems include data compression and speech compression.

Data compression is a process that is used encoding information so that fewer data bits of information are required to represent a given amount of data. Compression allows the transmission of more data over a given amount of time and circuit capacity. It also reduces the amount of memory required for data storage.

There are many types of data compression processes and their effectiveness depends on the underlying data. A simple data compression process replaces long sequences of bits with codes that take up a smaller amount of space. This can effectively compress large amounts of data that have long sequences of 1s or 0s with an instruction to create a long sequence and the value that will be used.

Figure 1.7 shows how a block of repetitive data can be substantially compressed using run length coding. This example shows a long string of 1s (such as in a black image). The RLC process begins with identifying the RLC code (a unique sequence or location in a data file), the character that will be repeated and the number of times that the character (or string of characters) will be repeated.

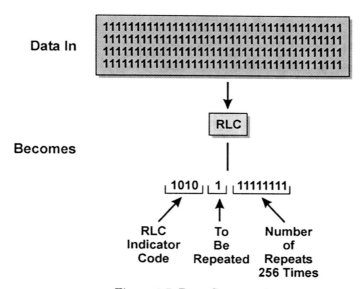

Figure 1.7, Data Compression

Speech compression (also called speech coding or voice coding) is a data compression device that characterizes and compresses digital speech information. A speech coder is also called a Coder/Decoder (CoDec) analyzes the incoming digital speech signal and converts the underlying speech sound it into a mathematical representation of that sound. Speech compression devices commonly use a table (a code book) that contains a cross-reference between specific sounds or speech patterns and code words that represent those sounds.

Figure 1.8 shows the basic digital speech compression process. In this example, the word "HELLO" is digitized. The initial digitized bits represent every specific shape of the digitized word HELLO. This digital information is analyzed and it is determined that this entire word can be represented by three sounds: "HeH" + "LeL" + "OH." Each of these sounds only requires a few digital bits instead of the many bits required to recreate the entire analog waveform.

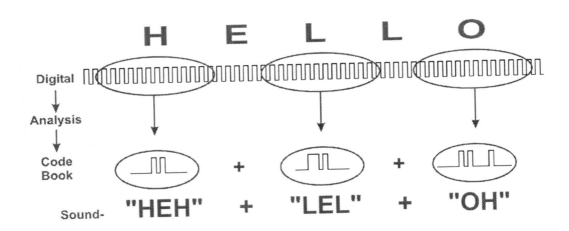

Figure 1.8, Speech Compression

Access Multiplexing

Access multiplexing is a process used by a communications system to coordinate and allow more than one user to access the communication channels within the system. There are four basic access multiplexing technologies used in wireless systems: frequency division multiple access (FDMA), time division multiple access (TDMA), code division multiple access, (CDMA), and space division multiple access (SDMA). Other forms of access multiplexing (such as voice activity multiplexing) use the fundamentals of these access-multiplexing technologies to operate.

FDMA systems use a process of allowing mobile radios to share radio frequency allocation by dividing up that allocation into separate radio channels where each radio device can communicate on a single radio channel during communication.

Figure 1.9 shows how a frequency band can be divided into several communication channels using frequency division multiplexing (FDM). When a device is communicating on a FDM system using a frequency carrier signal, its carrier channel is completely occupied by the transmission of the device.

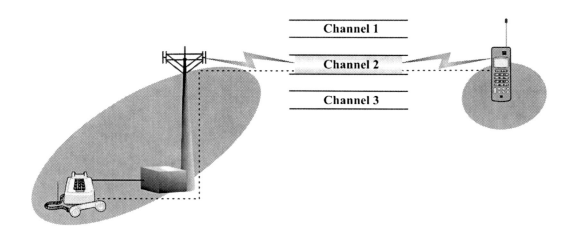

Figure 1.9, Frequency Division Multiple Access (FDMA)

For some FDM systems, after it has stopped transmitting, other transceivers may be assigned to that carrier channel frequency. When this process of assigning channels is organized, it is called frequency division multiple access (FDMA). Transceivers in an FDM system typically have the ability to tune to several different carrier channel frequencies.

TDMA systems allow several users to share a single radio channel by dividing the channel into time slots. When a mobile radio communicates with a TDMA system, it is assigned a specific time position on the radio channel. By allow several users to use different time positions (time slots) on a single radio channel, TDMA systems increase their ability to serve multiple users with a limited number of radio channels.

Figure 1.10 shows how a single carrier channel is time-sliced into three communication channels. Transceiver number 1 is communicating on time slot number 1 and mobile radio number 2 is communicating on time slot number 3. Each frame on this communication system has three time slots.

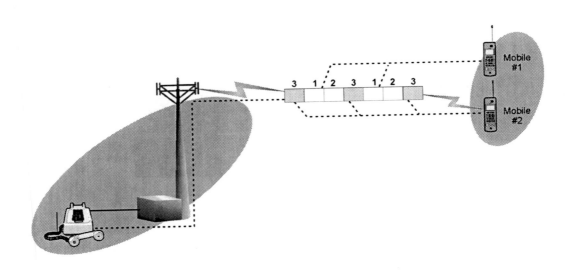

Figure 1.10, Time Division Multiple Access (TDMA)

Code division multiple access (CDMA), a form of spread spectrum communication. CDMA is a method of spreading information signals (typically digital signals) so the frequency bandwidth of the radio channel is much larger than the original information bandwidth.

Figure 1.11 shows how radio channels can provide multiple communication channels through the use of multiple coded channels. This diagram shows that a code pattern mask is used to decode each communication channel. The channel mask is shifted along the radio channel until the code chips (or a majority of the code chips) match the expected code pattern. When a match occurs, this produces a single bit of information (a logical 1 or 0). This example shows that the use of multiple code patterns (multiple masks in this example) allow multiple users to share the same radio channel.

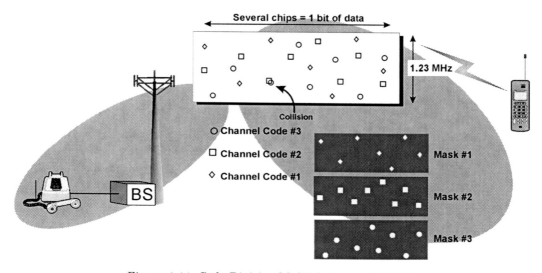

Figure 1.11, Code Division Multiple Access (CDMA)

Spatial division multiple access (SDMA) is a system access technology that allows a single transmitter location to provide multiple communication channels by dividing the radio coverage into focused radio beams that reuse the same frequency. To allow multiple accesses, each mobile radio is assigned to a focused radio beam. These radio beams may dynamically change with the location of the mobile radio. SDMA technology has been successfully used in satellite communications for several years.

Figure 1.12 shows an example of an SDMA system. Diagram (a) shows the conventional sectored method for communicating from a cell site to a mobile telephone. This system transmits a specific frequency to a defined (sectored) geographic area. Diagram (b) shows a top vies of a cell site that uses SDMA technology that is communicating with multiple mobile telephones operating within the same geographic area on a single frequency. In the SDMA system, multiple directional antennas or a phased array antenna system directs independent radio beams to different directions. As the mobile telephone moves within the sector, the system either switches to an alternate beam (for a multi-beam system) or adjusts the beam to the new direction (in an adaptive system).

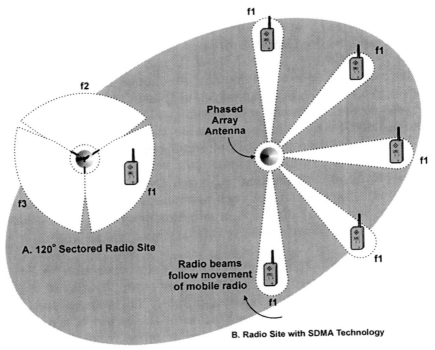

Figure 1.12, Spatial Division Multiple Access (SDMA)

Systems

Wireless systems are typically defined by their primary application. Because wireless systems are evolving into digital communication systems, this is beginning to blur the definition of wireless system type. For example, mobile telephone systems can also used for data and video services. Digital television systems can be used for Internet browsing and paging systems can be used for two-way data communications.

Mobile Telephone Systems (MTS)

Mobile telephone service (MTS) is a type of service where mobile radio telephones connect people to the public switched telephone system (PSTN) or to other mobile telephones. Mobile telephone service includes cellular, PCS, specialized and enhanced mobile radio, air-to-ground, marine, and railroad telephone services. MTS was a name used for the first public mobile telephone system used in the United States.

Mobile telephone systems allow mobile telephones to communicate with each other or to the public telephone system through an interconnected network of radio towers. In early mobile radio-telephone systems, one high-power transmitter served a large geographic area with a limited number of radio channels. Because each radio channel requires a certain frequency bandwidth (radio spectrum) and there is a very limited amount of radio spectrum available, this dramatically limited the number of radio channels that kept the serving capacity of such systems low. For example, in 1976, New York City had only 12 radio channels to support 545 customers and a two-year long waiting list of typically 3,700 [i].

One of the first mobile telephone systems began operation in St. Louis, Missouri, United States in 1946. By 1947, more than 25 cities in the United States had mobile telephone service available. The systems used a single high-power transmitter for the base station in the center of a metropolitan area. Coverage was provided for 50 miles or more from the transmitter. These initial systems used a human operator at the base station to manually connect the mobile user with the landline network. In most of these systems, service was very poor because too many customers (called subscribers) shared each radio channel (called loading). It was not uncommon to have busy channels over 50% of the time. Despite this poor service, it revolutionized the definition of telephone service and priority was given to police and ambulance service. The waiting list for mobile phones in some cities was more than 7 years. This type of system was improved many times and the last upgrade, called improved mobile telephone service (IMTS), was introduced in the mid 1960's. While there may still be some original systems in operation throughout the United States, new equipment for these systems is not currently being produced. It has been replaced with cellular systems.

When linked together to cover an entire metro area, the radio coverage areas (called cells) form a cellular structure resembling that of a honeycomb. The cellular systems are designed to have overlap at each cell boarder to enable a "hand-off" (also called a "handover") from one cell to the next. As a customer (called a subscriber) moves through a cellular or PCS system, the mobile switching center (MSC) coordinates and transfers calls from one cell to another and maintains call continuity.

Figure 1.13 shows a mobile telephone system. The wireless network connects mobile radios to each other or the public switched telephone network

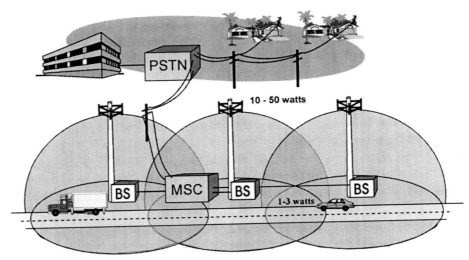

Figure 1.13, Mobile Telephone System

(PSTN) by using radio towers (base stations) that are connected to a mobile switching center (MSC). The mobile switching center can transfer calls to the PSTN.

When a cellular system is first established, it can effectively serve only a limited number of callers. When that limit is exceeded, callers experience too many system busy signals (known as blocking) and their calls cannot be completed. More callers can be served by adding more cells with smaller coverage areas - that is, by cell splitting. The increased number of smaller cells provides more available radio channels in a given area because it allows radio channels to be reused at closer geographical distances.

There are two basic types of systems: analog and digital. Analog systems typically use FM modulation to transfer voice information and digital systems use some form of phase modulation to transfer digital voice and data information. Although analog systems are capable of providing many of the services that digital systems offer, digital systems offer added flexibility as many of the features can be created by software changes. The trend at the end of the 1990's was for analog systems to convert to digital systems.

To allow the conversion from analog systems to digital systems, some cellular technologies allow for the use of dual-mode or multi-mode mobile telephones. These telephones are capable of operating on an analog or digital radio channel, depending on availability. Most dual-mode phones prefer to use digital radio channels in the event both are available. This allows them to take advantage of the new features such as short messaging and digital voice quality.

Cellular systems have several key differences that include the radio channel bandwidth, access technology type (FDMA, TDMA, CDMA or SDMA), data signaling rates of their control channel(s), and power levels. Analog cellular systems have very narrow radio channels that vary from 10 kHz to 30 kHz. Digital systems channel bandwidth ranges from 30 kHz to 5 MHz (or more). Access technologies determine how mobile telephones obtain service and how they share each radio channel. The data signaling rates determine how fast messages can be sent on control channels. The RF power level of mobile telephones and how the power level is controlled typically determines how far away the mobile telephone can operate from the base station (radio tower).

Cellular systems have evolved from narrowband analog mobile systems to multimedia wide band digital systems. Cellular systems are commonly categorized by generation numbers 1st generation (1G) through 4th generation (4G).

1st generation cellular systems (1G) are analog cellular systems. Analog cellular systems primarily provide voice service to a single user for each radio channel. While each radio tower may have multiple radio channels, the capacity of analog cellular systems is typically limited to 50 simultaneous communication channels per radio cover area.

2nd generation cellular systems (2G) are digital cellular systems. Digital cellular systems typically provide digital voice services to multiple users for each radio channel. The capacity of 2nd generation cellular systems was typically 3 to 20 times higher than for analog systems.

2.5 generation is a type of cellular systems that has more capabilities than digital cellular systems (2G) but has less capability than wideband multimedia cellular systems (3G). These systems provided advanced features that 2G systems could not offer such as packet data and various levels of quality of service (QoS).

3rd generation cellular systems (3G) are wideband multimedia communication systems. The requirements for 3G mobile communication systems are defined in the International Mobile Telecommunications "IMT-2000" project developed by the International Telecommunication Union (ITU). The IMT-2000 project that defined requirements for high-speed data transmission, Internet Protocol (IP)-based services, global roaming, and multimedia communications.

3G systems commonly use digital radio channels that are much wider (150 times wider) than analog cellular systems. This allows them to provide service to many more users per radio channel (50 to 100 users per channel) and permits these systems to offer multiple types of media services such as voice, data and video.

As of 2006, the requirements for fourth generation (4G) wireless networks are being defined. There is no single global vision for 4G as yet but the next generation of network is likely to be all IP-based, offering local (in building) data rates up to 1 Gbps, wide area data rates of 100 Mbps (rural) and support global mobility.

Figure 1.14 shows the evolution of mobile telephone systems. This figure shows that the first generation of cellular systems (1G) were analog, provided service to 1 user per radio channel and was primarily used for voice services. 2nd generation digital cellular (2G) systems use digitized radio channels that can be simultaneously shared by 3 to 20 users. 2.5 G systems added packet data capability to digital cellular systems for medium speed data services (e.g. Internet browsing). 3rd generation digital cellular (3G) have wide channels that were designed with the capability to provide multimedia broadband services and each 3G radio channel can be shared by approximately 50 to 100 users.

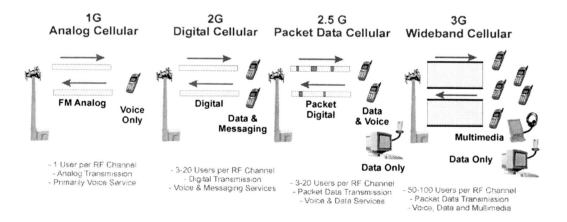

Figure 1.14, Evolution of Mobile Telephone Systems

Broadcast Radio

Radio broadcasting is the transmission of audio material (called a program) to a geographic area that is intended for general reception by the public. Broadcast radio service is typically paid for by advertising messages that are sent between audio programs.

Amplitude modulation (AM) radio broadcast services have been available for the past 100 years. Most AM radio broadcast systems use relatively low radio frequencies and very narrow radio channel bandwidth to efficiently deliver audio information over large geographic areas. Unfortunately, low frequency used for AM transmission often result in signals that sometimes skip long distances (hundreds of kilometers). This has the potential for interference in distant cities. Amplitude modulation is also easily subject to electrical noise and signal distortion. Recent advancements in AM modulation can allow channel coding for stereo and more reliable (less distorted) radio signals.

To overcome some of the limitations of AM, frequency modulation (FM) was developed. FM transmission is less susceptible to noise and distortion. Unfortunately, most FM broadcast systems use a wider radio channel than AM systems. FM broadcast channels can be up to 20 times the bandwidth of a single AM broadcast channel. The latest advancements in FM broadcasting include conversion from analog to digital and the ability to simultaneously send some additional information (sub-channels) with their audio broadcasts.

The current technology used for FM radio channel broadcast uses less bandwidth than is authorized for transmission. With some modifications to the transmitter, it has been possible for FM broadcast stations to simultaneously send some additional information (sub-channels) with their audio broadcasts. These sub-channels can contain audio or digital information. Sub-channels can be used for data transmission and paging services.

Figure 1.15 shows a typical radio broadcast system. The radio broadcast system consists of a production studio, a high-power AM or FM transmitter, a communications link between the studio and the transmitter, and network feeds for programming. Radio broadcasting involves the use of various types of information sources called "program sources." These program sources come from compact discs, tape recordings, soundproof audio studios, remote location sites (such as a van), or other network sources. The production studio controls and mixes the sources of information including audio compact discs, audio studio, audiotape, and other audio sources. A high-power transmitter broadcasts a single radio channel. The studio is connected to the transmitter by a coaxial cable, special leased telephone line (extra high quality), or dedicated radio link. Many radio broadcast stations receive their programming source from a radio broadcast network. This allows a single audio source to be relayed to many radio broadcast transmitters. The diagram also shows how a sub-channel is combined to provide a private audio broadcast service.

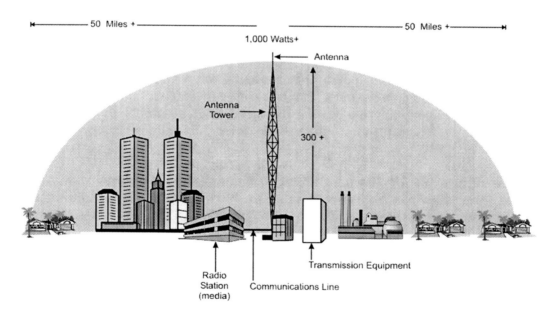

Figure 1.15, Radio Broadcast System

Two separate technologies are being deployed to offer digital audio broadcasting services. The first incorporates digital data into the conventional FM broadcast by adding the digital data signal to audio broadcasted signals. Devices that can receive these signals may receive the analog or digital signals. The second is a fully digital audio broadcasting (DAB) transmission that replaces conventional broadcasting systems.

Digital audio broadcasting (DAB) transmits voice and other information using digital radio transmission. DAB systems use a relatively wide digital radio channel that allows multiple communication channels (broadcast channels) to share a single radio channel.

The use of a single radio channel has several key benefits including reducing the transmission cost for each of the broadcasters, improving audio quality and reducing the effects of radio signal fading. DAB systems use a single radio transmission system that is shared by multiple audio broadcasters. This allows the cost of the transmission equipment and operations to be shared by multiple companies. DAB signals include error protection that allows for the correction of information that is lost or distorted during transmission. The frequency bandwidth of DAB radio signals is much wider than traditional FM radio channel bandwidths and this reduces the effects of signal fading. Overall, the audio quality of DAB signals tends to be more consistent than for FM broadcasted signals.

Development of the DAB system occurred in the 1990s and the selection of the audio coding and compression process was based on the MPEG audio coding, level 2 (MP2). MP2 audio coding is relatively efficient having relatively good quality at 128 kbps. While this digital audio signal is efficient, use of a limited bandwidth may result in a lower audio fidelity for DAB signals than for traditional FM signals in good signal quality conditions.

Figure 1.16 shows a digital audio broadcasting (DAB) system. This diagram shows that a DAB system uses a single wide digital radio channel that is divided into multiple digital audio channels. Audio broadcasters are linked to the DAB system by a digital connection and they are assigned (mapped) to a specific portion (logical channel) on the digital radio channel. The DAB system combines (multiplexes) the incoming channels to form one digital transmission channel. DAB receivers receive and decode the single frequency, separates out (demultiplexes) the specific digital (logical) channel and converts the digital channel back into its original audio form.

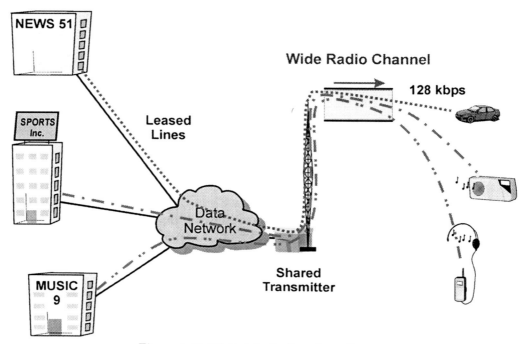

Figure 1.16, Digital Audio Broadcast System

Broadcast Television

Television broadcasting is the transmission of video and audio to a geographic area that is intended for general reception by the public, funded by commercials or government agencies. Land based television broadcasters (terrestrial broadcasters) transmit at high power levels from several hundred foot high towers. A high-power television broadcast station can reach over 50 miles.

Television broadcast systems have been evolving from analog broadcasting systems to digital multimedia broadcasting systems. Digital television broadcast systems have the capability of providing signals in various resolution formats ranging from low resolution for mobile devices to high definition television (HDTV). Analog television system standards include NTSC and PAL.

The standard television system used in the Americas is the National Television Standards Committee (NTSC) system. The first version of this system used 6 MHz RF channels to provide black and white television. The NTSC standard was later modified to allow color television signals to co-exist on the same type of video channel. The television system used in Europe and other parts of the world is phase alternating line (PAL). The PAL television system was developed in the 1980's to provide a common television standard in Europe. The PAL system uses 7 or 8 MHz wide radio channels.

Several enhancements have been added to this basic analog television broadcasting systems including audio stereo sound, closed captioning (very low data rate digital transfer) and ghost canceling.

Figure 1.17 shows an analog television broadcast system. This television system consists of a television production studio, a high-power transmitter, a communications link between the studio and the transmitter, and network feeds for programming. The production studio controls and mixes the sources of information including videotapes, video studio, computer created images (such as captions), and other video sources. A high-power transmitter broadcasts a single television channel. The television studio is connected to the transmitter by a high bandwidth communications link that can pass video and control signals. This communications link may be a wired (coax) line or a microwave link. Many television stations receive their video source from a television network. This allows a single video source to be relayed to many television transmitters.

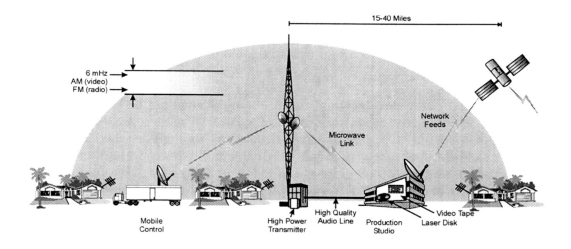

Figure 1.17, Television Broadcast System

Digital terrestrial television is the broadcasting of digital television signals using surface based (terrestrial) antennas. DTT systems transmit video, audio and other information using digital radio transmission. DTT systems use a relatively wide digital radio channel that allows multiple television and other media channels (broadcast channels) to share a single digital radio channel.

Each DTT radio channel can carry multiple television programs. A DTT system may be shared by multiple broadcasters allowing the cost of the transmission equipment and operations to be shared by multiple companies.

Development of the DTT system occurred in the 1990s and the selection of the video coding and compression process is typically based on the MPEG-2 coding standard. MPEG-2 video coding is relatively efficient having relatively good quality video at 2 to 4 Mbps.

There are several industry standards for DTT systems including digital video broadcasting (DVB), Integrated Services Digital Broadcasting (ISDB) and Digital Multimedia Broadcasting (DMB). These industry standards specify the radio channel characteristics including frequency, bandwidth, modulation types, error coding and compression.

These television broadcast industry standards may be used on or by other types of communication systems such as satellite, mobile television or cable television. When these standards are used for terrestrial (land based) television broadcasting, their acronym is typically followed or preceded by a letter that identifies the type of system (e.g. DVB-T or T-DMB is digital television over a terrestrial system).

Figure 1.18 shows a digital terrestrial television system. This diagram shows that a DTT system uses a single wide digital radio channel that is divided into multiple digital television channels. Television broadcasters are linked to the DTT system by a digital channel and they are assigned

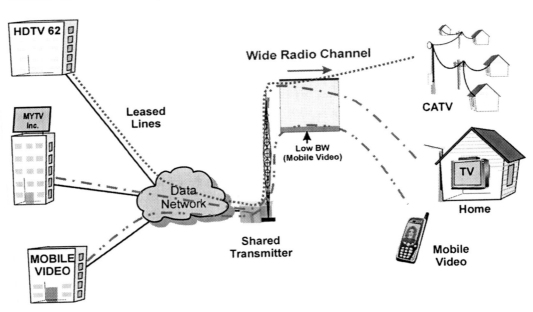

Figure 1.18, Digital Terrestrial Television (DTT) System

(mapped) to a specific portion (logical channel) on the digital radio channel. The DTT system combines (multiplexes) the incoming channels to form one digital transmission channel. DTT receivers receive and decode a DTT frequency, separate out (demultiplex) the specific digital (logical) channel and converts the digital channel back into its original television (video and audio) form.

Paging

Paging is a method of delivering a message, via a public or private communications system or radio signal, to a person whose exact whereabouts are unknown. Users as a rule carry a small paging receiver that displays a numeric or alphanumeric message displayed on an electronic readout or it could be sent and received as a voice message or other data.

Commercial paging service began in 1949 with the allocation of frequencies exclusively dedicated to one-way signaling services. Subscribers used AM receivers, listened for an operator to announce their number, and then called the service to receive their messages. Selective addressing (the ability to choose one individual pager from the group) was introduced in the mid 1950's and FM was first used in an experimental paging system in 1960. Pagers with alphanumeric displays made their debut in the early 1990's. In addition to complete messages that can be sent and stored in these pagers, a number of other services such as stock market and sports score reporting have been developed.

There are 4 basic types of messaging services offered by paging systems: tone, numeric, text (alpha), and voice. Two types of paging systems can deliver these messaging services: one-way and two-way paging. One-way paging systems only allow the sending of messages from the system to the pager. Two-way paging systems allow the confirmation and response of a message from the pager to the system as well.

One-way paging is a process where paging messages (signals) are sent from a radio tower to a pager without a return verification signal. In its simplest form, a one-way paging system can serve up to several hundred thousand numeric paging customers.

Figure 1.19 shows a one-way paging system. In this diagram, a high-power transmitter broadcasts a paging message to a relatively large geographic area. All pagers that operate on this system listen to all the pages sent, paying close attention for their specific address message. Paging messages are received and processed by a paging center. The paging center receives pages from the local telephone company or it may receive messages from a satellite network. After it receives these messages, they are sent after processing to the high-power paging transmitter by an encoder. The encoder converts the pagers telephone number or identification code entered by the caller to the necessary tones or digital signal to be sent by the paging transmitter.

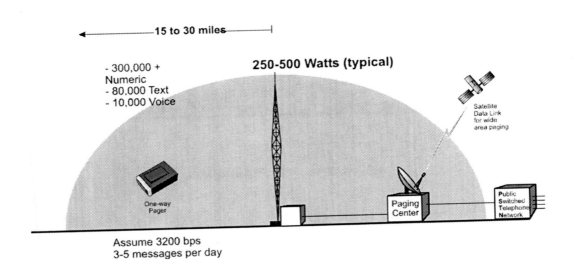

Figure 1., One-Way Paging System

Two-way paging systems allow the paging device to acknowledge and some-times respond to messages sent by a nearby paging tower. The two-way pager's low-power transmitter necessitates many receiving antennas being located close together to receive the low-power signal.

Figure 1.20 shows a high-power transmitter (200-500 Watts), which broad-casts a paging message to a relatively large geographic area and several receiving antennas. The reason for having multiple receiving antennas is that the transmit power level of pagers are much lower than the transmit power level of the paging radio tower. The receiving antennas are very sen-sitive, capable of receiving the signal from pagers transmitting only 1 watt.

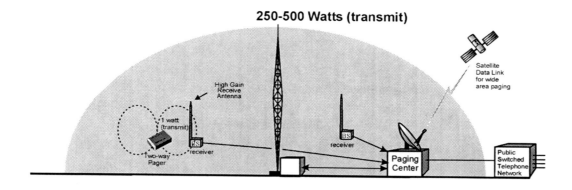

Figure 1.20, Two-Way Paging System

The number of required receivers for a two-way paging system is dependent on the available transmittal power from the paging and how fast the infor-mation is to be transferred. The higher the data transmission rate results in a higher number of required receivers.

The main advantage of two-way paging systems is their ability to require pagers to register their location within the paging system. This allows the paging system to direct pages for a specific pager only to the area near where the pager last registered. This frees up the paging capacity of channels in other geographic areas so paging messages can be sent to other pagers. This is a type of frequency reuse based on geographically separated systems.

Mobile Data Networks (MDN)

Mobile data networks (MDN), also called wireless wide area networks (WWANs) are wireless data transmission systems that cover large geographic area using cellular or public packet radio systems. Until the late 1990s, these systems were typically limited to data transfer rates below 20 kbps.

Initially, MDNs were primarily specialized networks that were only used for data communication. Through the 1980s and 1990s, various types of wireless networks such as FM broadcast, paging systems, and mobile telephone systems were adapted to provide data communication in addition to their originally designed service.

Some of the standardized dedicated MDNs include Mobitex and DataTAC. Mobitex is a packet data technology that was developed by Ericsson. Mobitex uses a frequency reuse system that allows for frequency handoff in a similar process that cellular systems perform. Mobitex systems typically operate in the 80 MHz, 400 MHz, and 800 MHz, 900 MHz frequency bands and Mobitex radios can transmit at a rate of 8 kbps. The DataTAC system uses the advanced radio data information service (ARDIS) technology that was developed by Motorola to provide public packet data service. ARDIS technology originated from the private network dedicated to IBM's service personnel in the 1980's. ARDIS commercial service started in January 1990 in the United States. The original ARDIS technology used the MDC4800 protocol that had a data transfer rate of 4,800 bps. This has evolved to use the RD-LAP protocol that allows a data transfer rate of 19,200 bps. The DataTAC Network uses single-frequency re-use (multiple base stations to the same receiver) to ensure good in-building coverage. ARDIS' wireless data network uses licensed radio channels (Specialized Mobile Radio Service) that usually operate in the 800 MHz frequency band.

FM broadcast radios provided slow-speed digital signals in addition to their audio signals. FM High Speed Sub-Band Signaling systems send data through FM radio channels in combination with other audio broadcast information. There are several radio stations that offer FM sub-band communication. The data rates of FM range from 1200 bps to over 10 kbps.

The first generation of mobile telephone systems (analog systems) used data modems to provide low-speed data service. 2nd generation mobile telephone systems (digital cellular) used simple data transfer adapters (DTAs) to provide low-speed data services. Towards the late 1990s, digital mobile telephone networks (2 ½ generation) began to offer medium-speed data services (up to 470 kbps.) Mobile telephone systems continue to evolve into high-speed wireless networks. The requirements for 3rd generation mobile telephone systems includes the ability to provide data transmission rates ranging from 144 kbps in all radio coverage areas (wide areas) up to 2 Mbps in urban areas (local areas).

Most wireless data services are dedicated to specific types of applications. Vertical wireless data applications (vertical) are very specific solutions, and have continued to win over mass-market "horizontal" offerings. Vertical solutions include applications such as utility meter reading or mobile dispatch. Horizontal solutions have mass-market appeal such as wireless e-mail.

Figure 1.21 shows a basic mobile wireless data system. In this system, many types of wireless data devices (mobile data terminals) communicate to nearby base station transmitters using a radio protocol that is unique to this system. A packet switch is used to route the packets between mobile data terminals or through a gateway that connects the mobile data system to the Internet. The packet switch is connected to a subscriber database that is used to determine which services are authorized, where the customer is located (location register), and the amount of services that the customer has used. This diagram shows that base station transmitters typically provide more transmitter power (50 Watts in this example) than the mobile data terminals (2 Watts in this example.).

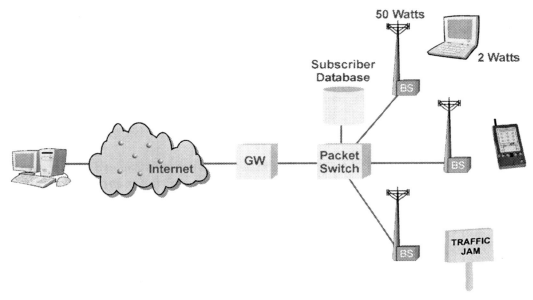

Figure 1.21, Mobile Data System

Land Mobile Radio (LMR)

Land mobile radio (LMR) systems are traditionally private systems that allow communication between a base and several mobile radios. LMR systems can share a single frequency or use dual frequencies. Land mobile radio may refer to a wide variety of mobile radio systems ranging from a simple pair of handheld "walkie-talkies" to digital cellular-like systems. LMR includes radio service between mobile units or between mobile units and a base station.

LMR systems that use a single frequency (called a simplex system); mobile radios must wait to talk when others are using the radio channel. To simplify the mobile radio design and increase system efficiency, some LMR systems use two frequencies; one for transmitting and another for receiving. If the radio cannot transmit and receive at the same time, the system is called half duplex. When LMR systems use two frequencies and can transmit and receive at the same time, this is called full duplex. When a company operates an LMR system to provide service to multiple users on a subscription basis (typically to companies), it is called a public land mobile radio system (PLMR).

Figure 1.22 shows a traditional two-way radio system. In this example, a high power base station (called a "base") is used to communicate with portable two-way radios. The two-way portable radios can communicate with the base or they can communicate directly with each other.

LMR systems are used by: taxicab companies, conventioneers, police and fire departments, and places where general dispatching for service is a normal course of business communications. SMR radios are regularly designed to be rugged to survive the harsh environment. SMR radios can usually be programmed with a unique code. This code may be an individual code or

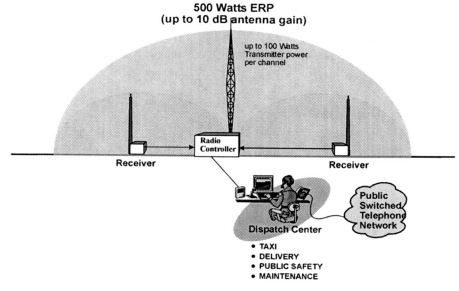

Figure 1.22, Traditional Land Mobile Radio System

group code (e.g., pre-designated group of users such as a fire department). This allows all the radios belonging to a group, or a sub-group, to be "paged" by any party in the group. A push-to-talk method is used during the dispatch call (page) or reply. This push-to-talk radio-to-radio communication efficiently utilizes the airwaves because of the bursty (very short transmission time) nature of the information.

Trunked radio is a mobile communication system that allows mobile radios (called trunked radios) to access more than one of the available radio channels in that system. The radio access control procedure allows the mobile to access alternate radio channels in the system in the event that the channel it has requested is busy.

Figure 1.23 shows a typical trunked SMR system. In this example, there are several available radio channels. Mobile radios that operate in this system that wish to communicate may search for an available radio channel by looking for identification tones. Optionally, some trunked radio systems use dedicated radio channels to coordinate access to radio channels.

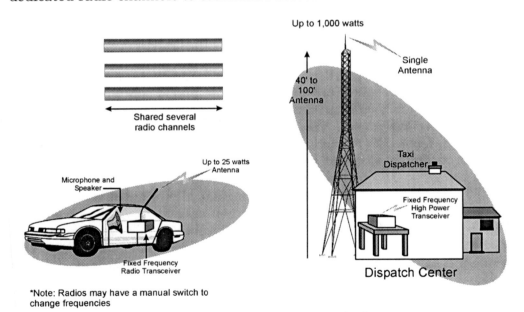

Figure 1.23, Trunked Land Mobile Radio System

Automated land mobile radio systems are divided into two categories; SMR or Enhanced SMR (ESMR). Enhanced land mobile radio systems operate and have similar features to mobile telephone systems.

Land mobile radio systems are evolving from analog transmission (e.g. AM or FM) to digital land mobile radio (DLMR) communication systems. DLMR systems allow communication between a base and several mobile radios using digital modulation technology. DLMR systems can share a single frequency or use dual frequencies. DLMR services commonly include various types of voice and data services for police, taxi, fire and other types of dispatch services. Some of the common DLMR systems include Tetra and iDEN.

TETRA is a digital land mobile radio system that was formerly called Trans European Trunked Radio. The TETRA was developed by the European Telecommunications Standards Institute (ETSI) to create a more efficient and flexible communication services from both private and public-access mobile radio users.

TETRA is capable of sending and receiving short data messages simultaneously with an ongoing speech call. It effectively supports voice groups and has capacity for over 16 million identities per network (over 16 thousand networks per country). TETRA permits direct mode operation (talk around) that permits direct communication between mobile radios without the network. TETRA includes a priority feature to help guarantee access to the network by emergency users. The system allows independent allocation of uplinks and downlinks to increases system efficiency. The signaling protocol supports sleep modes that increase the battery life in mobile radios.

The TETRA system is fully digital system that allows for mixed voice and data communication. It is specified in open standards. The TETRA system allows up to 4 users to share each 25 kHz channel. It allows inter-working with other communication networks via standard interfaces. TETRA is capable of call handoff between cells and it has integrated security (user/network authentication, air-interface encryption, end-to-end encryption).

Integrated dispatch enhanced network is a digital radio system that provides for voice, dispatch and data services. iDEN was formerly called Motorola integrated radio system (MIRS). iDEN was deployed in 1996 for enhanced specialized mobile radio (E-SMR) service. The iDEN system radio channel bandwidth is 25 kHz and it is divided into frames that have 6 time slots per frame. The iDEN system allows 6 mobile radios to simultaneously share a single radio channel for dispatch voice quality and up to 3 mobile radios can simultaneously share a radio channel for cellular like voice quality.

Figure 1.24 shows a typical digital two-way radio system. In this example, a digital mobile is connected to a data display in a mobile vehicle. The radio transceiver has a transmitter and receiver bundled together. Some digital systems combine the control channel with a traffic channel while others use dedicated control channels. Digital systems can either use a single digital radio channel for each user or can divide a single radio channel into time slots for use. The use of time division allows for several users to simultaneously share each radio channel.

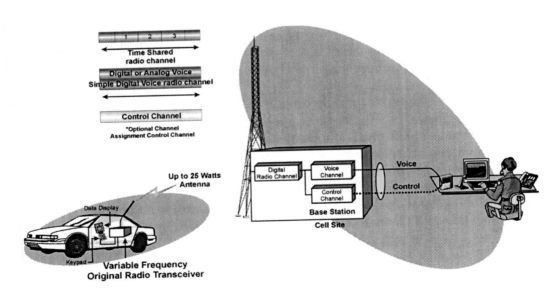

Figure 1.24, Digital Land Mobile Radio System

Aircraft Telephones

Aircraft telephones allow people on an airplane to initiate telephone calls with the public telephone system through connection via land based radio or satellite transmission systems. Recently, some aircraft telephone systems have been upgraded to allow calls to be received on the airplane.

Aircraft telephone systems are ordinarily a hybrid wireless system that is a terrestrial wireless system (land-based) combined with satellite service. The terrestrial system is used to connect telephone calls when the aircraft is above land and is within distance of a ground transmitter. For the terrestrial-based system, the phone handset in the airplane is connected to a transmitter in the plane's belly that connects the call down to one of the ground antennas located strategically throughout the country. The call is routed to a ground switching station that connects the call to the receiving party.

The satellite system is used mainly over the water, where calls are out of reach of the ground antennas. For the satellite-based system, the phone handset on the plane is connected to an antenna on the top of the plane that connects the signal up to an orbiting satellite. The call is then sent down to earth by the satellite frequencies to its satellite earth station, then to one of the main ground switching stations that routes the call to the PSTN.

Aircraft phone systems normally have handsets in a common area or handsets that are located in the back of passenger seats. If the handset is located in the seat, some aircraft phone systems allow incoming calls. For someone to reach you on an aviation telephone system, the person on the aircraft must first get a telephone access number and temporary identification code by registering with the aviation telephone operator. The person placing the call from the ground dials the access number and enters the temporary identification code and the call will be routed to the aviation telephone.

Figure 1.25 shows a public aircraft telephone system. This diagram shows that aircraft may be served by terrestrial (land-based) systems or satellite communication systems. In either case, the aircraft communicates with a gateway that links the radio system to the public telephone system.

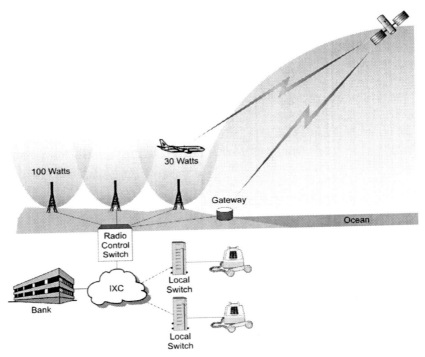

Figure 1.25, Public Aircraft Telephone System

Satellite

Satellite communication systems use orbiting satellites to relay communications signals from one satellite station to one or several other users. Satellite communication can be divided into categories of fixed satellite service, positioning systems, and mobile satellite communication systems.

There are three basic types of satellite systems: geosynchronous earth orbit (GEO), medium earth orbit (MEO), and low earth orbit (LEO). GEO satellites hover at approximately 22,300 miles above the surface of the earth. GEO satellites revolve along with the earth once a day; they appear stationary with respect to the earth. The high-gain antennas used to receive signals from 22 thousand miles away (usually called "dish" antennas) are pointed directly toward the satellite. MEO satellites are located closer to the earth than GEO satellites and do not as a rule require high-gain antennas. This is important as MEO satellites revolve around the earth several times per day and fixed antennas cannot be used. The newest satellite technology being deployed is LEO satellites. LEO satellites are located approximately 450 miles above the surface of the earth. Because these satellites are relatively close to the earth, portable phones with smaller antennas can be used.

Figure 1.26 shows the different types of satellite communication systems. The GEO satellite system is primarily used for television broadcast services,

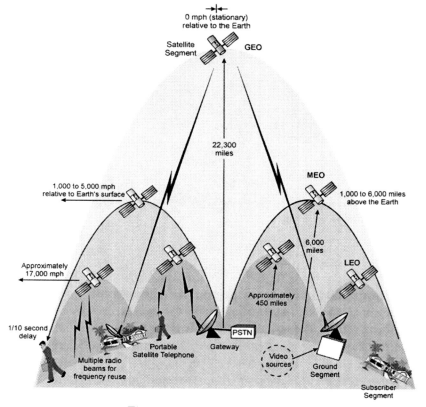

Figure 1.26, Satellite Systems

as their satellites appear stationary above the Earth. MEO and LEO systems are used for mobile communications as they are located much closer to the Earth. However, these satellites continuously move relative to the surface of the Earth.

Mobile satellite telephone service allows customers to use specialized satellite mobile telephones to communicate in any part of the world to the PSTN through the use of communication satellites. Commercial communication satellite services began in the mid-1960's with the establishment of Intelsat, a multinational organization with well over 130 member nations today. An organization known as the Communications Satellite Corporation (COMSAT) also was established in the early-1960's and became the United States' representative in Intelsat. These first commercial applications of satellites provided international telephone and television program transmission, primarily between the United States and Europe.

Wireless PBX (WPBX)

Wireless private branch exchange (wireless office) telephone systems are used in a business environment to provide similar features as a private branch exchange (PBX) system with the added capability of providing mobility throughout the office area. The wireless office commonly begins with a specialized private branch exchange (PBX) system that that has been adapted to allow connection to wireless devices (portable office telephones). While more complex than a home cordless telephone, it is not typically as complex as a complete cellular telephone system.

The WPBX telephone radio coverage area is usually within one or more company buildings or on a campus. The more popular WPBX systems use unlicensed frequencies with a protocol available only to the manufacturer of the WPBX. Ordinarily, WPBX telephones cannot be used outside the established campus. These private WPBX systems use small wall mounted antennas, and like cellular, the space is divided to provide adequate capacity for the expected usage. A popular WPBX system is digital enhanced cordless telecommunication (DECT).

The DECT system is a digital cordless and WPBX system. DECT was originally developed by the European Telecommunications Standards Institute (ETSI) technical standards committee in the late 1980s and the specification was released in 1992 and commercial equipment was available by 1993. The number of DECT handsets in use by 2004 is in excess of 50 million. It was first intended that the use of the DECT system be for wireless office. After its release, it has been adapted to allow home cordless, public cordless, and radio local loop (RLL).

The DECT system includes three key parts; the mobile radio portable part (PP), the radio base station fixed part (RFP), and the interconnecting system fixed part (FP). There are two version of DECT; the European version and the American version. The European version uses a very wide radio channel to allow up to 12 simultaneous wireless telephones to share each channel. The American version uses a slightly more narrow radio channel and allows up to 8 users to share a single radio channel. Personal Wireless Telecommunication (PWT) is an adaptation of DECT for the North American market.

DECT technology is managed and promoted by the DECT forum. The DECT forum helps to promote DECT technology worldwide, assists in the allocation of radio frequencies for DECT systems provides forums that allows developers and providers to share information, and to manage the evolution of DECT technology to ensure reasonable migration from older legacy equipment to improved versions of DECT. More information about DECT forum and technology can be found at www.DECT.org.

Figure 1.27 shows a sample WPBX radio system. A WPBX system typically has a switching system that is located at the company. The WPBX switch interfaces a PSTN communication line and multiple radio base stations. Radio base stations communicate with wireless office telephones that can move throughout the system. A control terminal is used to configure and update the WPBX with information about the wireless office telephones and how they can be connected to the PSTN.

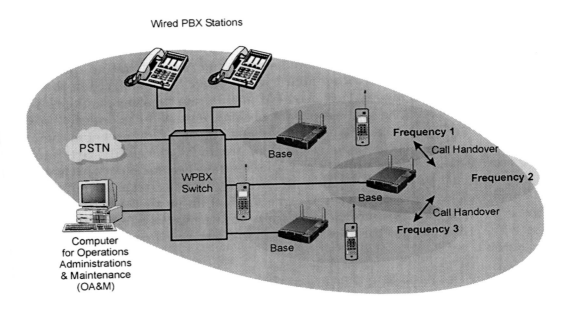

Wired PBX Stations

Figure 1.27, Wireless Office Telephone System

Recent hybrids have been developed whereby the telephone handset has two technologies built into the operation of the phone. When the telephone is inside the WPBX coverage area (preferred) it acts as a private phone; when outside the WPBX coverage area, the phone has the ability to send and receive calls on the public cellular system, incurring airtime charges as any other cellular user.

Residential Cordless

Cordless systems are short-range wireless telephone systems that are primarily used in residential applications. Cordless telephones regularly use radio transmitters that have a maximum power level below 10 milliWatts (0.01 Watts). This limits their usable range to 100 meters or less.

The earliest generation of home cordless telephones used a single radio channel that used amplitude modulation. These first generation cordless phones were susceptible to electrical noise (static) from various types of electronic equipment such as florescent lights. The noise encountered when using these phones sometimes created a consumer impression that cordless telephone quality was below standard wired telephone quality. Improved versions of cordless phones that used FM modulation to overcome the electrical noise resulted. As cordless phones became more popular, interference from nearby phones became a problem. In apartment buildings where there were many users of cordless phones in close proximity, the ability to initiate and receive calls could be difficult as radio channels became busy with many users. This led to the development of cordless phones that used multiple radio channels. As voice privacy became more of an issue, cordless phones began to use scrambled voice. Some of these voice privacy systems were analog while a majority of cordless phones that offer voice privacy use digital transmission.

Figure 1.28 shows the evolution of cordless telephones. Until the mid 1990's, most cordless telephones were limited to use in a small radio coverage area of their base station that was usually located in the home. That home base station was normally connected to the telephone line of the owner (either residential or a single office telephone line) and they were *not* intended to serve the general public. To add more value to the use of cordless phones, cordless telephones evolved to allow access to base stations in public locations. Cordless telephones could then be used in the home and in areas that were served by public base stations. The next evolution for cordless telephones was the combination of other types of wireless products and services into the cordless phone. This included the combination of wireless office and cellular telephones into a cordless phone.

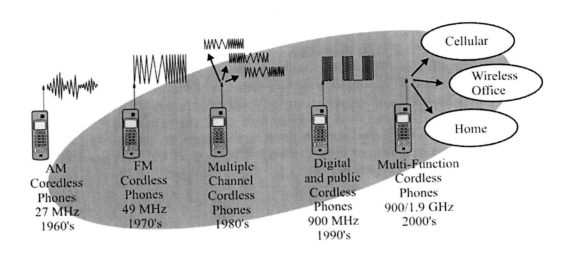

Figure 1.28, Evolution of Cordless Telephone Systems

Most home cordless telephones used frequencies in unlicensed radio frequency bands. Because so many homes operate cordless phones, each manufacturer must build-in circuitry to minimize the interference caused by other cordless devices. The original cordless phones use a very crowded frequency band (around 27 and 49 MHz) utilizing analog radio wave modulation. Recently, cordless telephones have been developed that operate in the 902-928 MHz or 2.4 GHz unlicensed industrial, scientific, and medical (ISM) frequency band.

Residential cordless telephones must automatically coordinate their radio channel access as they operate independently of any type of network control. To coordinate radio channel access and avoid interference to other cordless handsets installed in the vicinity, cordless phones perform radio channel scanning and interference detecting prior to transmitting a signal.

Because cordless telephone systems do not as a rule have a dedicated control channel to provide information, the cordless handset and base station continuously scan all of the available channels (typically 10 to 25 channels).

Figure 1.29 shows the basic cordless telephone coordination process. This diagram shows that when the cordless phone or base station desires to transmit, the unit will choose an unused radio channel and begin to transmit a pilot tone or digital code with a unique identification code to indicate a request for service. The other cordless device (base station or cordless phone) will detect this request for service when it is scanning and its receiver will stop scanning and transmit an acknowledgement to the request for service. After both devices have communicated, conversation can begin. When another nearby base station detects the request for service, it will determine that the message is not intended for it and will not process the call and scanning will continue.

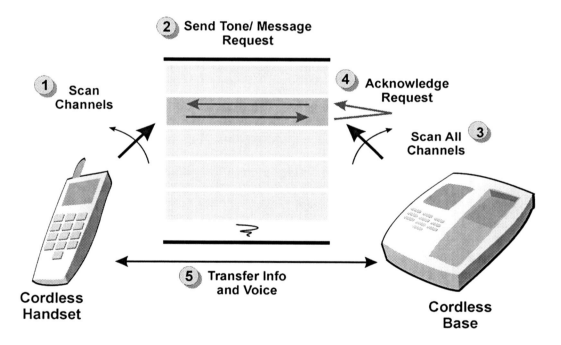

Figure 1.29, Cordless Telephone System

Wireless Personal Area Network (WPAN)

Wireless personal area networks (WPANs) are temporary (ad-hoc) short-range wireless communication systems that typically connect personal accessories such as headsets, keyboards, and portable devices to communications equipment and networks.

WPANs usually operate in unlicensed (uncontrolled) frequency bands. While there are many unlicensed frequency bands, the frequency band of 2.4 GHz to 2.483 GHz is popular for WPAN networks as it is available in most countries throughout the world.

The devices in WPAN systems use very low radio transmission power of 1 milliwatt to 100 milliwatts (1/10th of a Watt.) For unlicensed use, radio transmission is authorized for all users provided the radio equipment conforms to unlicensed transmission requirements. Anyone can use the unlicensed frequency band but there is no guarantee they will perform at peak performance due to possible interference. Some types of WPAN systems include Bluetooth and UWB.

Bluetooth is a wireless personal area network (WPAN) communication system standard that allows for wireless data connections to be dynamically added and removed between nearby devices. Each Bluetooth wireless network can contain up to 8 active devices and is called a Piconet. Piconets can be linked through each other to form Scatternets. The system control for Bluetooth requires one device to operate as the coordinating device (a master) and all the other devices are slaves. This is very similar to the structure of a universal serial bus (USB). However, unlike USB connections, most Bluetooth devices can operate as either a master (coordinator) or slave and Bluetooth devices can reverse their roles if necessary.

Some of the common types of devices that can be linked by wireless personal area network communication include a computer and all of its accessories. Devices such as a keyboard, mouse, display, speakers, microphone, and a presentation projector. As these devices are brought within a few feet of each other, they automatically discover the availability and capabilities of other devices. If these devices have been setup to allow communication with other devices, the user will be able to use these devices as if they were directly connected with each other. As the devices are removed from the area or turned off, the option to use these devices will be disabled from the user.

Figure 1.30 shows how a Bluetooth system can connect multiple devices on a single radio link. This diagram shows that a laptop computer has requested a data file from a desktop computer. When this laptop computer first

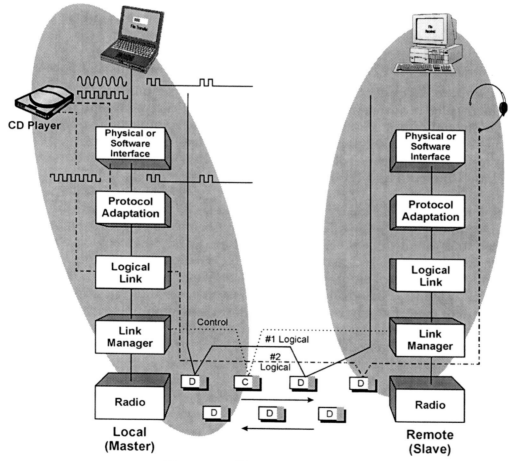

Figure 1.30, Bluetooth Operation

requests the data file, it accessed the Bluetooth radio through a serial data communication port. The serial data port was adapted to Bluetooth protocol (RFComm) and a physical radio channel was requested from the local device (master) to the remote computing device (slave). The link manager of the master Bluetooth device requests a physical link to the remote Bluetooth radio. After the physical link is created, the logical link controller sends a message to the remote device requesting a logical channel be connected between the laptop computer and the remote computer. The logical link continually transmits data between the devices. In this diagram, the user then requests that a CD player send digital audio to a headset at the remote computer. Because a physical channel is already established, the logical link controller only needs to setup a 2nd logical link between the master Bluetooth device and the remote Bluetooth device. Now data from the CD ROM will be routed over the same physical link between the two Bluetooth devices.

The UWB system transmits information at very high data transmission rates over very short distances. Ultra wideband is a method of transmission that transmits information over a much wider bandwidth (perhaps several GHz) than is required to transmit the information signal. Because the UWB signal energy is distributed over a very wide frequency range, the interference it causes to other signals operating within the UWB frequency band is extremely small. This may allow the simultaneous operation of UWB transmitters and other existing communication systems with almost undetectable interference.

Wireless Local Area Network (WLAN)

A wireless local area network (WLAN) allows computers and workstations to communicate with each other using radio propagation as the transmission medium. The wireless LAN can be connected to an existing wired LAN as an extension, or can form the basis of a new network. While adaptable to both indoor and outdoor environments, wireless LANs are especially suited to indoor locations such as office buildings, manufacturing floors, hospitals and universities.

Wireless LAN's generally use either infrared or radio frequency (RF) as their transmission media. Infrared is line-of-sight only, and poses problems in many office environments when viewed as a single solution. When coupled with twisted pair wire (the basic LAN media) and used to bring in isolated workstations across a factory floor, it has proven to be a more reliable technology. RF is not line-of-sight and thus is not subject to the problems of infrared. It does, however, encounter interference from many devices found in the office and factory.

Wireless LANs often used radio channels in an unlicensed frequency band. These wireless data systems can transmit data up to 54 Mbps on a single channel (2-11 Mbps is more typical). Inexpensive WLAN systems can also be setup as point-to-point wireless data systems through the use of directional antennas. This setup would allow a WLAN system to interconnect data networks between buildings within a campus up to approximately 25 miles. Providing this wireless data link only requires the installation of 2 antennas with a clear line of site communication.

Wireless LAN systems typically use the unlicensed radio frequency bands instrument, scientific and medial (ISM) frequency bands. These bands include 902-928 MHz, 2.4 - 2.483 GHz, and 5.7 GHz ranges with the most common unlicensed frequency band used throughout the world being the 2.4 GHz frequency band.

WLANs typically operate up to a distance of 300 feet (100 meters). WLAN systems provide much larger coverage by interconnected radio access nodes. Wireless LAN standards include multiple versions of IEEE 802.11 and HyperLAN.

A WLAN system typically includes radio access ports and extension ports. The extension ports shown in the figure are PCMCIA cards that plug into a laptop computer. These extension ports communicate via radio-to-radio access ports. The radio access ports convert the WLAN radio signal back into computer network signals (such as Ethernet or token ring).

Radio transmission in WLAN systems may be coordinated (centralized) or independent (distributed). When one of the devices in the WLAN system coordinates the network, it is called point coordination function (PCF).

When devices are allowed to independently communication with each other, it is called distributed control function (DCF).

PCF allows the coordinated operation (assigned access control - Infrastructure Mode) of wireless data devices (stations). In PCF contention free system, communication devices wait until they receive a polling message before they transmit any information. Because a master host coordinates the transmission of all the devices within its networks, no device will transmit at the same time (contention free). The access control portion of a PCF session usually involves requiring the communication agreeing to listen to a single host before transmitting any data. To confirm transmitted data has been successfully received; the polling message will usually include information about the status of packets that have been received. If the sending device does not receive a confirmation of transmission in the polling message, it will retransmit the data again after it receives another polling message.

Figure 1.31 shows a wireless LAN system that uses a point control function (PCF) that requires each of the access devices to wait until they hear a poll

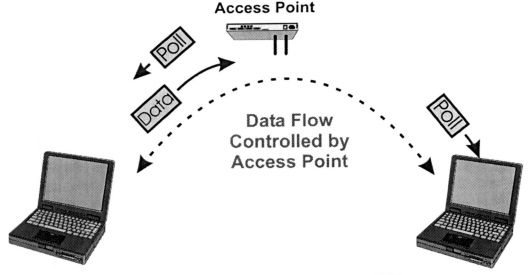

Figure 1.31, WLAN Point Control Function (PCF)

request before they can transmit. When a station receives a token from a serving station (such as an access point), it may respond to the poll request with the information it has to transmit. The use of the PCF mode ensures that channel collisions do not occur. This allows the serving station a guarantee a specific data transfer (data flow) to specific stations. This guaranteed data transmission rate is important for real time communications (such as voice or video).

DCF allows the independent operation (distributed access control) of wireless data devices (stations). In DCF contention-based system, communication devices randomly request service from channels within a communication system. Because communication requests occur randomly, two or more communication devices may request service simultaneously. The access control portion of a DCF session usually involves requiring the communication device to sense for activity before transmitting and listen for message collisions after its service request. If the requesting device does not hear a response to its request, it will wait a random amount of time before repeating the access attempt. The amount of random time waited between retransmission requests increases each time a collision occurs.

DCF mode is a peer-to-peer network where the temporarily wireless network that has no server or central access point, hub, or router. Since there is no central base station to monitor traffic or provide Internet access, the various signals can collide with each other.

Figure 1.32 shows how a distributed WLAN system allows units to independently request and transmit data. Using a carrier sense multiple access (CSMA) protocol, WLAN data terminals listen for activity in the radio channel before starting to transmit. If there is no channel activity, the WLAN data terminal can begin to transmit. It will then listen to hear a response to its transmitted signal. If it does not hear a response within a pre-defined time period, it will stop transmitting and start the channel access process again (listen then transmit).

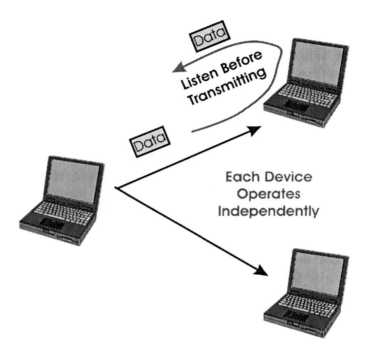

Figure 1.32, WLAN Distributed Access Control

WLAN devices that use PCF and DCF may co-exist with each other as DCF devices will not transmit during PCF transmission periods.

Wireless Cable

Wireless Cable is the common term assigned to a radio frequency-based alternative to the cable TV distribution system. An example of wireless cable technologies is multichannel multipoint distribution system (MMDS) or local multichannel distribution system (LMDS). By 1998, there were over 10 million wireless customers throughout the world and over 1.1 million in the United States.

Wireless cable system can simultaneously supply local television channels and high-speed data services. MMDS has a high-speed standardized air interface allowing mass deployment of cable television service by the new unregulated telephone companies. Cable television providers, who have access to most homes, can now provide telephone service.

In 1996, some analog MMDS systems began upgrading to digital service. Through the use of digital video compression, digital transmission allows 5 or 6 times the video channel capacity. In addition to video programming, wireless cable can provide telephone service and data services.

Figure 1.33 shows that the major component of a wireless cable system is the head-end equipment. The head-end equipment is equivalent to a telephone central office. The head-end building has a satellite connection for cable channels and video players for video on demand. The head-end is linked to base stations (BS) which transmit radio frequency signals for

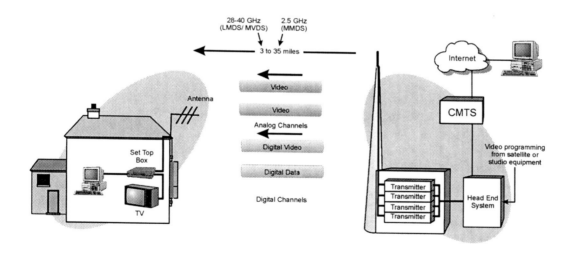

Figure 1.33, Wireless Cable System

reception. An antenna and receiver in the home converts the microwave radio signals into the standard television channels for use in the home. Like traditional cable systems a set-top box decodes the signal for input to the television. Low frequency wireless cable systems can reach up to approximately 70 miles.

Wireless cable is one of the most economical technologies available for the delivery of pay television service. Wireless cable systems do not require extensive networks of cables and amplifiers, bringing the offered price generally lower than a traditional cable service. To the customer, a wireless cable system operates in the same manner as a traditional cable system. Because wireless signals are transmitted over the air rather than through underground or above-ground cable networks, wireless systems may be less susceptible to outages, offer better signal quality and be less expensive to operate and maintain than traditional cable systems. In conventional coaxial cable distribution networks, the television signal quality declines in strength as it travels along the cables and must be boosted by amplifiers thus introducing distortion into the television signal.

To add security for wireless cable systems, so unauthorized users do not gain access to the system (stealing service), signals from video sources are scrambled with a code. The user must have the code to successfully view the video signals. Like traditional cable systems, wireless cable systems employ "addressable" subscriber authorization technology, which enables the system operator to control centrally the programming available to each individual subscriber, such as a pay-per-view selection

There are two primary methods of providing a communication path back from the end customer to the network operator: a telephone line and wireless. Wireless cable systems have typically only provided wireless downlink service (radio transmission from the system to the customer). Some of the new wireless cable systems now dedicate some of their radio channel capacity to uplink channels (from the customer to the system). Uplink channels allow the customer to select programming sources (such as pay per view) or may allow two-way Internet access.

Wireless Local Loop (WLL)

Wireless local loop (WLL) service refers to the distribution of telephone service from the nearest telephone central office to individual customers via a wireless link. In some cases, it is referred to as "the last mile" in a telephone network. This term is a bit misleading, though, because the coverage area of a WLL system may extend several miles from switching facilities.

Competitive local exchange carriers (CLEC) are competitors to the incumbent local exchange carriers (ILECS) and are can use WLL systems to rapidly deploy competing systems. If CLECs do not use wireless systems, they must either pay the existing phone company for access to the local loop (resale) or dig and install their own wire to the local customers.

Figure 1.34 shows a wireless local loop system. In this diagram, a central office switch is connected via a fiberoptic cable to radio transmitters located

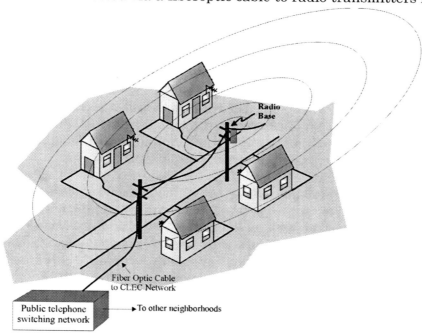

Figure 1.34, Wireless Local Loop

in a residential neighborhoods. Each house that desires to have dial tone service from the WLL service provider has a radio receiver mounted outside with a dial tone converter box. The dial tone converter box changes the radio signal into the dial tone that can be used in standard telephone devices such as answering machines and fax machines. It is also possible for the customer to have one or more wireless (cordless) telephones to use in the house and to use around the residential area where the WLL transmitters are located.

The most basic service offered by wireless local loop (WLL) system is to provide standard dial tone service known as plain old telephone service (POTS). In addition to the basic services, WLL systems typically offer advanced features such as high-speed data, residential area cordless service, and in some cases, video services. To add value to WLL systems, WLL service providers will likely integrate and bundle standard phone service with other services such as cellular, paging, high speed Internet, or cable service.

WLL systems can provide for single or multiple-line units that connect to one or more standard telephones. The telephone interface devices may include battery back up for use during power outages. Most wireless local loop (WLL) systems provide for both voice and data services. The available data rates for WLL systems vary from 9.6 kbps to over several hundred kbps. WLL systems can be provided on cellular and PCS, private mobile radio, unlicensed cordless, and proprietary wideband systems that operate the 3.4 GHz range.

Integrated access device converts multiple types of input signals into a common communications format. IADs can be used by WLL systems to integrate different types of telephone devices (e.g. phone line, Internet data connection and digital television) onto a common digital medium (broadband wireless connection).

Figure 1.35 shows an integrated access device (IAD) combines multiple types of media (voice, data, and video) onto one common data communications system. This diagram shows that three types of communication devices (telephone, television, and computer) can share one data line (e.g. DSL or Cable Modem) through an IAD. The IAD coordinates the logical channel assignment for device and provides the necessary conversion (interface) between the data signal and the device. In this example, the telephone interface provides a dialtone or ISDN signal and converts the dialed digits into messages that can be sent on the data channel. The video interface buffers and converts digital video into the necessary video format for the television or set top box. The data interface converts the line data signal into Ethernet (or other format) that can be used to communicate with the computer. This diagram also shows that the IAD must coordinate the bandwidth allocation so real time signals (such as voice) are transmitted in a precise scheduled format (isochronous). The digital television signal uses a varying amount of bandwidth as rapidly changing images require additional bandwidth. The IAD also allocates data transmission to the computer as the data transmission bandwidth becomes available (what is left after the voice and video applications use their bandwidth).

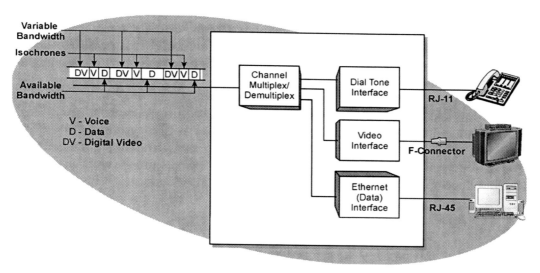

Figure 1.35, Integrated Wireless Local Loop Services

Worldwide Interoperability for Microwave Access (WiMax)

Worldwide Interoperability for Microwave Access (WiMax) is a wireless communication system that allows computers and workstations to connect to high-speed data networks (such as the Internet) using radio waves as the transmission medium with data transmission rates that can exceed 155 Mbps for each radio channel. The WiMax system is defined in a group of IEEE 802.16 industry standards and its various revisions are used for particular forms of fixed and mobile broadband wireless access.

WiMax is system that is primarily used as a wireless metropolitan area network (WMAN). WMANs can provide broadband data communication access throughout an urban or city geographic area. WMANs are used throughout the world and their applications include consumer broadband wireless Internet services, interconnecting lines (leased lines) and transport of digital television (IPTV) services. WiMax broadband wireless can compete with DSL, cable modem and optical broadband connections.

The 802.16 system was initially designed for fixed location or Nomadic service. Nomadic service is the providing of fixed communication services to more than one location. While nomadic service may be provided too many locations, nomadic service typically requires the transportable communication device to be fixed in location during the usage of communication service.

WiMax has several different physical radio transmission options which allow WiMax systems to be deployed in areas with different regulatory and frequency availability requirements. The WiMax system was designed with the ability to be used in licensed or unlicensed frequency bands using narrow or wide frequency channels.

Figure 1.36 shows some of the different types of uses that WiMax networks can provide. This diagram shows that WiMax systems can be used for point-to-point links, residential broadband or high-speed business connections. This example shows that the point to point (PTP) connection may be independent from all other systems or networks. The point to multipoint (PMP) system allows a radio system to provide services to multiple users. WiMax systems can also be setup as mesh networks allowing the WiMax system to forward packets between base stations and subscribers without having to install communication lines between base stations.

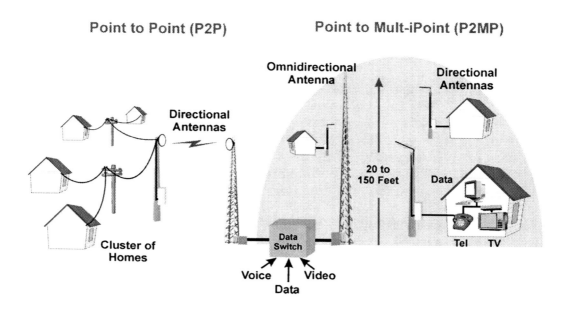

Point to Point (P2P)
Point to Mult-iPoint (P2MP)

Figure 1.36, WiMax System Types

Services

The key services for wireless communications include mobile voice, paging, wireless data, audio broadcast and television broadcast services.

Mobile Voice

Voice communication can be telephony; wide area (cellular), business location (wireless office) or home cordless (residential) or voice paging, dispatch (fleet coordination) or group voice (audio broadcasting). Service rates for voice applications typically involve an initial connection charge, basic monthly minimum fee, plus an airtime usage charge. When the customer uses service in another system than their home registered system, there may be a daily fee and/or a higher per minute usage fee.

Figure 1.37 shows that mobile telephone usage charges include recurring costs (monthly fee) and usage costs (per minute). Many mobile rate plans include some on-peak (during the business day) and off-peak (evening and weekend) minutes. When the user exceeds these included minutes, additional fees are charged per minute. For mobile telephones that have the capability to operate outside the existing system (Roam into other geographic areas) or on other networks (such as either analog or digital systems), higher roaming usage charges usually apply. Mobile telephone systems also offer advanced features. In many cases, the advanced features are offered for free (e.g., voice mail) as they increase the amount of usage.

Monthly Fee	$39.99
Peak Minutes Included	400
Off-Peak Minutes Included	3000
Additional Minutes	$0.40/minute
Long Distance Rate Per Minute	$0.15/minute
Roaming Rate	$0.60/minute
Other Services (Voice Mail, 3 Way Calling)	Typically Included

Figure 1.37, Mobile Telephone Usage Costs

Paging

Paging services include tone paging, numeric paging, alpha (text) paging, and voice paging. Tone paging service notifies a paging customer that a message has been sent via a tone. This tone usually is designated to mean a call-back to a single location is requested. The original "beep, beep" tone pagers that started it all have become very popular as entry-level private communications systems, including restaurants (beeping servers when orders are ready) and in other centralized locations where a tone page necessitates a choice in response to only one.

While the popularity of these types of pagers has decreased overall, some retailers continue to offer tone pagers as "loss leaders" allowing the publication of low prices in print ads to attract attention. Still, for dispatch operators or other industries that only need a response to a central office, tone pagers offer a cost-effective solution.

Numeric paging is the sending of paging messages, typically telephone numbers that are displayed on a small paging device. After the message is received, the user calls back the displayed telephone number to talk to the sender.

Figure 1.38 shows sample service rate plans for numeric, text, and voice paging services. This example shows that paging service typically involves signing up for local, regional, or nationwide coverage. Paging service rate plans usually have a monthly service charge that allows pages using a local telephone number. Additional charges may be assessed for 800 toll free and 0800 freephone access numbers. Paging carriers usually offer a maximum number of pages per month and pages that additional per-page charges may apply for pages that are sent beyond the pre-defined limit. This chart also shows that the cost of numeric paging is typically lower than text (alpha) paging and that text paging is typically less than voice paging. There are usually additional fees for operator services such as receiving calls and converting the calls into text messages.

	Numeric	Text (Alpha)	Voice
Activation Fee	$15	$15	$15
Monthly Access Fee (includes 500 pages)	$7.50	$15	$25
Additional Pages	$0.10 each	$0.10	$0.10
Regional Coverage	Add $3.00 per month	Add $6.00 per month	Limited Availability
National Coverage	Add $6.00 per month	Add $12.00 per month	Limited Availability
Voice Operator Text Message	N/A	$0.50 per message	N/A

Figure 1.38, Paging Cost

Alpha paging displays numerical and textual information. This can take many forms, from verbatim text messages to weather reports. Messages can be recorded into voice mail where operators type them up and send them out, or callers can dictate messages to live operators directly. Many carriers bundle free news, weather and sports feeds from other sources. While some carriers offer unlimited numeric usage, many charge for additional pages (both numeric and text message) beyond a pre-defined limit. Some carriers charge so much per character while others simply charge by the message.

Voice pagers broadcast messages through a built-in speaker in the paging unit. Message volume settings can usually be set to loud, soft or private, through an earpiece.

Two-way pagers allow fully interactive capabilities, permitting users to respond to pages by sending an original or canned alphanumeric message. Additional hardware must be purchased (or leased). Customers can lease the paging device for about $15 per month or purchase it for around $399.00.

Wireless Data

There are three basic services offered by wireless data systems: circuit switched data, packet switched data and messaging.

Circuit switched data is a bearer service as it only transports the user's data between points. When sending data through a circuit switched connection, the user regularly pays a standard per-minute charge for the amount of time that the connection is maintained regardless of how much data is sent through the channel.

It usually takes several seconds to establish a circuit switched connection on a wireless network. This is due to the processing of dialed digits through the telephone network and the amount of time the modem requires to establish which communication language will be used (called training time). The user ordinarily pays for this setup time even if they only have a very small amount of information to send (such as an email message). Once a connection is established on a circuit switched connection, data transfer rates generally range from 9600 bps up to 28,800 bps.

Packet switched data is also a bearer type of service as it only transports the users data between points. When sending packets of data through the network, the user normally pays only for the amount of data or number of packets that they send.

Unlike circuit switched data, the connection time for packets is ordinarily under 1 second (some systems may be below 150 msec) and the user does not pay for this setup time. The typical price for packet data transmission ranges from approximately less than 1 cent to $1 per Megabyte. A one-time activation fee is as a rule required along with a minimum monthly fee. The usage amount is normally applied to the monthly fee.

Several wireless data service providers in the United States now offer service based on application and number of units. This results in different price plans that can vary from $15-25 per month per unit with some systems offering a flat fee for a fixed or unlimited amount data transmission. There is usually a monthly recurring fee that provides the user with a monthly amount of data. If the customer uses the maximum data allocation, an additional fee per kilobyte or megabyte of data is charge.

Wireless messaging is a teleservice as it processes the user data. Wireless messaging services include store, forward, and Internet connectivity. Typically wireless messaging is combined (bundled) with wireless data service (such as packet data).

Figure 1.39 shows that the cost of sending mobile data has dropped from over $100 per megabyte (10 cents per kilobyte) in the early 1990s to under $1 per megabyte rate plan. This diagram also shows that the cost of packet data (short bursts of data) has typically been higher than circuit switched (continuous) data. However, in the early 2000s, the cost of packet switched data is approaching the cost of circuit switched data.

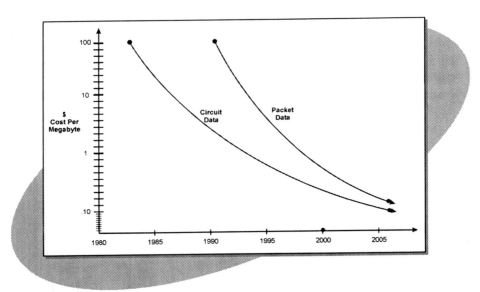

Figure 1.39, Mobile Data Cost

Broadcast Radio

Audio broadcast services are the transmission of program material (typically audio) that is typically paid for by advertising. Most commercial stations receive the bulk of their ad revenues from local advertising, as opposed to television, which gets most of its revenue from network advertising.

According to the radio advertising bureau (RAB), in the United States, broadcast radio is received by approximately 94% of consumers listen to radio and a majority of consumers (95%) spend more than 22 3/4 hours listening to radio each week. [2].

Broadcast radio revenues are primarily obtained from the providing of advertising. A majority of the advertising revenue for radio broadcasting comes from local (spot) advertising. In the United States, According to the National Association of Broadcasters in 2005, total broadcast advertising radio revenues reached over $20 billion [3].

Figure 1.40 shows the growth of the broadcast radio advertising industry in the United States between 1998 and 2005. This chart shows that total broadcast advertising revenue has increased from $15.4 billion in 1998 to more than $21 billion in 2005. This graph shows how radio broadcast advertising revenue is divided between local, national and network advertising. This example shows that local spot advertising in 2005 was 78% of total radio broadcast advertising revenue.

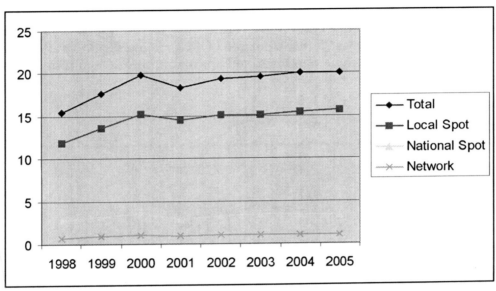

Figure 1.40, Radio Advertising Revenue 1998 to 2005 in the United States
Source: Radio Advertising Bureau (www.RAB.com)

Broadcast Television

Television broadcast services are the transmission of television program material (typically video combined with audio) that is typically paid for by advertising. Most television stations receive a substantial amount of ad revenues from network advertising, as opposed to radio, which gets most of its revenue from local advertising.

According to the television bureau of advertising (TVB), television advertising revenue in the United States was approximately $68 billion and a majority of consumers spend between 3 and 4 hours watching television each day [4].

Advertising revenue for television broadcasting comes from network, spot ad, local, syndicates and cable advertising. According to the TVB, the percentage of advertising expenditures has been shifting from network advertising to local and cable advertising [5].

Figure 1.41 shows that television advertising revenue in the United States has increased from approximately $49 billion in 1998 to almost $68 billion in 2005. This chart shows that the portion of advertising spent on television as compared to other media has remained approximately 25% of total advertising expenditures.

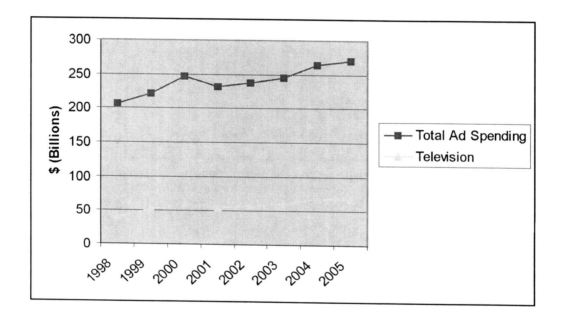

Figure 1.41, Television Advertising Revenue 1998 to 2005 in the United States
Source: Television Bureau of Advertising (www.TVB.org)

References:

1. GSM World, World Statistics, www.GSMWorld.com.
2. "2006 Radio Marketing Guide and Fact Book," New York, NY, 2006, www.RAB.Org.
3. "2006 Radio Marketing Guide and Fact Book," New York, NY, 2006, Radio Advertising Bureau, www.RAB.Org, pg 23.
4. "2005 Television Ad Revenue Figures," New York, NY, 2006, Television Bureau of Advertising, www.TVB.Org.
5. "Trends in Television," New York, NY, 2006, Television Bureau of Advertising, www.TVB.Org.

Appendix 1

Acronyms

1G-First Generation
2.5G-Second And A Half Generation
2G-Second Generation
3G-Third Generation
4G-Fourth Generation
A-LAN-Appartment Local Area Network
AM-Amplitude Modulation
ARDIS-Advanced Radio Data Information Service
AuC-Authentication Center
BS-Base Station
CDMA-Code Division Multiple Access
Codec-Coder/Decoder
CSMA-Carrier Sense Multiple Access
DAB-Digital Audio Broadcast
DCF-Distributed Coordinated Function
DECT-Digital Enhanced Cordless Telephone
DLMR-Digital Land Mobile Radio
DMB-Digital Multimedia Broadcasting
DOC-Department Of Communications
DTA-Data Transfer Adapter
DTT-Digital Terrestrial Television

DVB-Digital Video Broadcast
ESMR-Enhanced Specialized Mobile Radio
ETSI-European Telecommunications Standards Institute
FCC-Federal Communications Commission
FDM-Frequency Division Multiplexing
FM-Frequency Modulation
FP-Fixed Part
GEO-Geosynchronous Earth Orbit
GPS-Global Positioning System
HLR-Home Location Register
IAD-Integrated Access Device
ILEC-Incumbent Local Exchange Carrier
IMTS-Improved Mobile Telephone Service
IP-Internet Protocol
IPTV-Internet Protocol Television
ISDB-Integrated Services Digital Broadcasting
ISM-Instrument, Scientific and Medical Band
ITU-International Telecommunication Union
LEO-Low Earth Orbit

LMDS-Local Multichannel Distribution Service
LMR-Land Mobile Radio
MDN-Mobile Data Network
MEO-Medium Earth Orbit
MIRS-Motorola Integrated Radio System
MMDS-Multichannel Multipoint Distribution Service
MP3-Motion Picture Experts Group Layer 3
MSC-Mobile Switching Center
NTSC-National Television System Committee
PBX-Private Branch Exchange
PCF-802.11 Point Coordination Function
PMP-Point to Multipoint
POTS-Plain Old Telephone Service
PP-Portable Part
QoS-Quality Of Service
RAB-Random Access Burst
RFCOMM-Radio Frequency Communication Port
RFP-Request For Proposal
RF-Radio Frequency
RLL-Radio Local Loop
USB-Universal Serial Bus
VLR-Visitor Location Register
WiMax-Worldwide Interoperability for Microwave Access
WLAN-Wireless Local Area Network
WLL-Wireless Local Loop
WMAN-Wireless Metropolitan-Area Network
WMM-Wi-Fi Multimedia

WPAN-Wireless Personal Area Network
WPBX-Wireless Private Branch Exchange

Index

ALTHOS

Althos Publishing Book List
2006-2007

Product ID	Title	# Pages	ISBN	Price	Copyright
Billing					
BK7781338	Billing Dictionary	644	1932813381	$39.99	2006
BK7781339	Creating RFPs for Billing Systems	94	193281339X	$19.99	2007
BK7781373	Introduction to IPTV Billing	60	193281373X	$14.99	2006
BK7781384	Introduction to Telecom Billing, 2nd Edition	68	1932813845	$19.99	2007
BK7781343	Introduction to Utility Billing	92	1932813438	$19.99	2007
BK7769438	Introduction to Wireless Billing	44	097469438X	$14.99	2004
IP Telephony					
BK7781311	Creating RFPs for IP Telephony Communication Systems	86	193281311X	$19.99	2004
BK7780530	Internet Telephone Basics	224	0972805303	$29.99	2003
BK7727877	Introduction to IP Telephony, 2nd Edition	112	0974278777	$19.99	2006
BK7780538	Introduction to SIP IP Telephony Systems	144	0972805389	$14.99	2003
BK7769430	Introduction to SS7 and IP	56	0974694304	$12.99	2004
BK7781309	IP Telephony Basics	324	1932813098	$34.99	2004
BK7781361	Tehrani's IP Telephony Dictionary, 2nd Edition	628	1932813616	$39.99	2005
BK7780532	Voice over Data Networks for Managers	348	097280532X	$49.99	2003
IP Television					
BK7781362	Creating RFPs for IP Television Systems	86	1932813624	$19.99	2007
BK7781355	Introduction to Data Multicasting	68	1932813551	$19.99	2006
BK7781340	Introduction to Digital Rights Management (DRM)	84	1932813403	$19.99	2006
BK7781351	Introduction to IP Audio	64	1932813519	$19.99	2006
BK7781335	Introduction to IP Television	104	1932813357	$19.99	2006
BK7781341	Introduction to IP Video	88	1932813411	$19.99	2006
BK7781352	Introduction to Mobile Video	68	1932813527	$19.99	2006
BK7781353	Introduction to MPEG	72	1932813535	$19.99	2006
BK7781342	Introduction to Premises Distribution Networks (PDN)	68	193281342X	$19.99	2006
BK7781357	IP Television Directory	154	1932813578	$89.99	2007
BK7781356	IPTV Basics	308	193281356X	$39.99	2006
BK7781389	IPTV Business Opportunities	232	1932813896	$24.99	2007
BK7781334	IPTV Dictionary	652	1932813349	$39.99	2006
Legal and Regulatory					
BK7781378	Not so Patently Obvious	224	1932813780	$39.99	2006
BK7780533	Patent or Perish	220	0972805338	$39.95	2003
BK7769433	Practical Patent Strategies Used by Successful Companies	48	0974694339	$14.99	2003
BK7781332	Strategic Patent Planning for Software Companies	58	1932813322	$14.99	2004
Telecom					
BK7781313	ATM Basics	156	1932813136	$29.99	2004
BK7781345	Introduction to Digital Subscriber Line (DSL)	72	1932813454	$14.99	2005
BK7727872	Introduction to Private Telephone Systems 2nd Edition	86	0974278726	$14.99	2005
BK7727876	Introduction to Public Switched Telephone 2nd Edition	54	0974278769	$14.99	2005
BK7781302	Introduction to SS7	138	1932813020	$19.99	2004
BK7781315	Introduction to Switching Systems	92	1932813152	$19.99	2007
BK7781314	Introduction to Telecom Signaling	88	1932813144	$19.99	2007
BK7727870	Introduction to Transmission Systems	52	097427870X	$14.99	2004
BK7780537	SS7 Basics, 3rd Edition	276	0972805370	$34.99	2003
BK7780535	Telecom Basics, 3rd Edition	354	0972805354	$29.99	2003
BK7781316	Telecom Dictionary	744	1932813160	$39.99	2006
BK7780539	Telecom Systems	384	0972805397	$39.99	2006

For a complete list please visit
www.AlthosBooks.com

Althos Publishing Book List
2006-2007

Product ID	Title	# Pages	ISBN	Price	Copyright
Wireless					
BK7769434	Introduction to 802.11 Wireless LAN (WLAN)	62	0974694347	$14.99	2004
BK7781374	Introduction to 802.16 WiMax	116	1932813748	$19.99	2006
BK7781307	Introduction to Analog Cellular	84	1932813071	$19.99	2006
BK7769435	Introduction to Bluetooth	60	0974694355	$14.99	2004
BK7781305	Introduction to Code Division Multiple Access (CDMA)	100	1932813055	$14.99	2004
BK7781308	Introduction to EVDO	84	193281308X	$14.99	2004
BK7781306	Introduction to GPRS and EDGE	98	1932813063	$14.99	2004
BK7781370	Introduction to Global Positioning System (GPS)	92	1932813705	$19.99	2007
BK7781304	Introduction to GSM	110	1932813047	$14.99	2004
BK7781391	Introduction to HSPDA	88	1932813918	$19.99	2007
BK7781390	Introduction to IP Multimedia Subsystem (IMS)	116	193281390X	$19.99	2006
BK7769439	Introduction to Mobile Data	62	0974694398	$14.99	2005
BK7769432	Introduction to Mobile Telephone Systems	48	0974694320	$10.99	2003
BK7769437	Introduction to Paging Systems	42	0974694371	$14.99	2004
BK7769436	Introduction to Private Land Mobile Radio	52	0974694363	$14.99	2004
BK7727878	Introduction to Satellite Systems	72	0974278785	$14.99	2005
BK7781312	Introduction to WCDMA	112	1932813128	$14.99	2004
BK7727879	Introduction to Wireless Systems, 2nd Edition	76	0974278793	$19.99	2006
BK7781337	Mobile Systems	468	1932813373	$39.99	2007
BK7769431	Wireless Dictionary	670	0974694312	$39.99	2005
BK7780534	Wireless Systems	536	0972805346	$34.99	2004
BK7781303	Wireless Technology Basics	50	1932813039	$12.99	2004
Optical					
BK7781386	Fiber Optic Basics	316	1932813861	$34.99	2006
BK7781329	Introduction to Optical Communication	132	1932813292	$14.99	2006
BK7781365	Optical Dictionary	712	1932813659	$39.99	2007
Marketing					
BK7781318	Introduction to eMail Marketing	88	1932813187	$19.99	2007
BK7781322	Introduction to Internet AdWord Marketing	92	1932813225	$19.99	2007
BK7781320	Introduction to Internet Affiliate Marketing	88	1932813209	$19.99	2007
BK7781317	Introduction to Internet Marketing	104	1932813292	$19.99	2006
BK7781317	Introduction to Search Engine Optimization (SEO)	84	1932813179	$19.99	2007
BK7781323	Web Marketing Dictionary	688	1932813233	$39.99	2007
Programming					
BK7781300	Introduction to xHTML:	58	1932813004	$14.99	2004
BK7727875	Wireless Markup Language (WML)	287	0974278750	$34.99	2003
Datacom					
BK7781331	Datacom Basics	324	1932813314	$39.99	2007
BK7781355	Introduction to Data Multicasting	104	1932813551	$19.99	
BK7727873	Introduction to Data Networks, 2nd Edition	64	0974278734	$19.99	2006
Cable Television					
BK7780536	Introduction to Cable Television, 2nd Edition	96	0972805362	$19.99	2006
BK7781380	Introduction to DOCSIS	104	1932813802	$19.99	2007
BK7781371	Cable Television Dictionary	628	1932813713	$39.99	2007
Business					
BK7781368	Career Coach	92	1932813683	$14.99	2006
BK7781359	How to Get Private Business Loans	56	1932813594	$14.99	2005
BK7781369	Sales Representative Agreements	96	1932813691	$19.99	2007
BK7781364	Efficient Selling	156	1932813640	$24.99	2007

For a complete list please visit
www.AlthosBooks.com